Rockies, for example, the oaks, junipers, and pinyon pines of the foothills give way at moderate elevations to ponderosa pine, which itself is replaced at higher altitudes by Douglas fir and then by spruce and alpine fir, until one finally emerges into the stark, treeless, oddly beautiful Alpine Zone, whose rigorous conditions have resulted in plants so tiny that they are best examined on hands and knees.

Emphasizing plant-animal associations and the general adaptations of animals to varying environments, Mr. Costello also writes of the mammals of the mountain world—the large ones, such as the elk, bighorn, and cougar, as well as their smaller relatives, including coyotes, rabbits, wolverine, shrews, and beaver. Here, too, are the amphibians, the reptiles, and the insects —the latter so often dismissed as pests although essential links in the cycle of life, as pollinators, and as converters of dead leaves, branches, and trees into substances that other animals and plants can use again.

A concluding chapter tells of mountains and man—of their early exploration by trappers, of their conquest by miners and by railroad-builders, and finally of the ways they are used—and sometimes abused—today by stockmen, loggers, miners, and the average citizen, camper, hunter, and snowmobiler. An appendix on "How to See the Mountains," with emphasis on safety precautions, a glossary of geological and ecological terms, a bibliography, and an index complement the text.

DAVID F. COSTELLO, a native of Nebraska, now lives in the shadow of the Rockies, in Fort Collins, Colorado. Recently retired, he worked for many years in research posts for the U.S. Forest Service.

The Mountain World

Hunting wild boar with dogs in the Santa Lucia Mountains in California. These rugged mountains rise tier on tier above the Big Sur near Carmel, California.

The
Mountain
World

David F. Costello

ILLUSTRATED BY THE AUTHOR

THOMAS Y. CROWELL COMPANY
Established 1834   New York

Manufactured in the United States of America

Library of Congress Cataloging in Publication Data

Costello, David Francis, date
The mountain world.

    Bibliography: p.
    Includes index.
    1. Mountain ecology. 2. Mountains. I. Title.
QH87.C67 1975 500.9'143   74-34369
ISBN 0-690-00695-0

1 2 3 4 5 6 7 8 9 10

To Cecilia

Books by David F. Costello

The Desert World

The Prairie World

The Mountain World

The World of the Prairie Dog

The World of the Ant

The World of the Gull

The World of the Porcupine

Contents

The Mountain World

The Gunnison River in Colorado has carved its way through granite and shale. The resulting spectacular land forms include the Black Canyon with vertical walls, pinnacles, rim-rock formations, talus slopes and verdant meadow lands.

1

The Mountain World

ONE OF THE FINEST EXPERIENCES of man is to climb a mountain. The approach from the flatness of the sea, the plains, or the desert always gives hope for surprise. If you begin your ascent on one of the eastern mountains you leave the disciplined green correctness of fields and conventional towns and villages and follow the course of a mountain stream. As the hills grow higher and the valley narrows, the river winds, and the trail writhes in sympathy. Farther up, the river tosses and tumbles in its bed. Its banks are modestly screened by trees. Now the wrinkles in the earth begin in earnest and quickly grow into high ridges. The river divides into lesser streams, each flowing from its own valley. You follow the ridges and at last you arrive on the massive shoulders of the world. There below you, if it is autumn, the landscape is mottled with green, russet, crimson, and red, and the forested peaks are close together and they leave no room for farms or villages.

If you approach one of the western mountains, the topography, scenery, and plant life are different. The Southern Rocky Mountains in Colorado, for example, appear as a great wall rising from 5,000 feet at the western edge of the Great Plains to cliffs and peaks that tower 14,000 feet or more into the clouds. At the edge of the mountain front you begin your ascent through canyons with tilted layers of sedimentary rocks where the earth's bare bones are exposed or only slightly covered with the flesh of soil. Above the foothills as the canyons open into valleys the water in creeks and rivers tosses and tumbles. Here and there are towns and settlements with houses lined up within arms-reach of each other because space is so limited.

Beyond the towns, as the valleys rise, pine forests on the slopes give way to spruce and fir. Still higher, the slopes recede and verdant meadows appear where sheep and cattle graze. And then, back of beyond, the landscape opens into high elevation grassy parklands covering hundreds of square miles. From the rims of these parks majestic mountains rise more thousands of feet, thrusting their treeless snow-covered summits far into the sky. If you are a sturdy mountaineer, rugged enough to conquer one of the peaks, the world lies at your feet and the view extends to far horizons where serrated ridges of other mountains mark the boundary between earth and sky.

There are thousands of mountains to be explored by those who love the challenge of all nature. The jagged ranges of the Great Basin rise brown and drab under the noonday sun. These mountains are rugged and their climate belies the heat of the desert far below. Trees grow on their slopes. And trout rise to flies in clear streams and in subalpine lakes watered by winter snows. From the peaks you see the substance of mountains, washed by ages of erosion into the desert flats far below while miles away other desert ranges set the limits of your visible world. But you know that beyond their peaks are still other ranges in bewildering succession until the mighty ramparts of the Sierra Nevada and the volcanic cones of the Cascades are reached.

On many a mountain top you can pick up a handful of gravel or a stone and see crystals or grains that were sorted by waves and layered into stone beneath ancient seas. On other mountains are fossils of fish, shells, and plants. They tell us that life existed aeons ago, and now that they lie thousands of feet high on a mountainside, they tell us that nothing is static on this earth.

The soaring mountains seem formidable and immutable, but time and the physical forces of nature are still at work upon them. In the northern Cascades, for example, in the tremendous setting of peaks and rounded valleys, glaciers are still grinding away the rocks of ages. They remind us that the continent once was half covered with an enormous ice sheet that lowered the ocean hundreds of feet. We are further reminded by the lava

Rock strata in the Uinta Mountains in Utah are strongly tilted but the
trees obey the law of gravity. The hard rocks have resisted erosion
while the soft layers have less successfully resisted weathering.

covered slopes of volcanic mountains in Washington, Oregon,
and northern California that convulsions of the earth have occurred in relatively recent times.

Mountains are more than piles of rocks which slowly change
through geologic time. Life in many forms exists on mountains
and affects even the form and the existence of the rocks of ages.
The landscape is mellowed by the vegetation growing on foothills, valleys, high slopes, and even the rocky terrain far above
the upper timberline. Vegetation with its associated animals exhibits a multitude of biotic assemblages which interact with the
physical-chemical environment. How these actions and reac-
tions occur in nature will be discussed in more detail in later
pages. But consider some examples here.

A lichen produces acid that etches the rock on which it grows and speeds the rock's disintegration together with the expansive forces of heat from the sun in summer and the cold of frost in winter. The lichen also produces organic matter in which mosses later grow and still later form a substratum for seed plants. Through succession, or the replacement of different combinations of plants through time, a forest may finally cover the place where lichens once grew on bare rocks.

The forest with its ground layer of herbs and shrubs holds the soil in place and slows erosion, thus retarding the rate at which the mountain erodes and its substance is carried to the sea. Likewise, the beaver, which depends on trees for its food and material for its house, dams the river to make its pond and thus slows the water that carries away the sediments and substances of mountains.

The biotic assemblages of plants and animals on mountains appear in zones of shrublands, forests, and alpine meadows as one ascends to ever higher altitudes. The boundaries between these zones are usually distinct and the transition from one to another is abrupt. The number of zones and their biotic composition vary among different mountain ranges, especially those in the West, with their proximity to the sea on their windward sides, and with their nearness to the desert on their leeward sides. On north-facing slopes, forests grow at lower elevations, while on south-facing slopes, shrublands occur higher on the mountainsides.

Altitude has much the same effect on vegetation as does latitude. If you ascend from the foothills in the Southern Rockies to the top of a 14,000 foot mountain the passage through vegetation zones is equivalent to a journey from southern Colorado to the tundra zone in Alaska. Each band of vegetation descends as one goes northward; even the alpine grasslands descend until they reach sea level in the Arctic regions.

In the mountains with their topographic diversity, many ecological habitats occur in the different zones. In the alpine region, local habitats vary from ponds and meadows, to sloping grasslands, gravel beds, and boulder fields. In the spruce-fir forests are openings where meadow plants flourish and willows and herbs border the streams. In this environment the mammals, birds, and insects differ from those in the nearby forest. Farther down are aspen groves and lodgepole pine forests which are products of fires that formerly denuded spruce and fir forests. Aspen grows quickly in burned areas because it regenerates from root sprouts. Lodgepole pine also revegetates, since fire opens its cones and results in abundant seed distribution. Spruces and firs are killed by fire, and the return to climax forest is extremely slow.

Aspen and lodgepole forests are temporary forests which in time are invaded and replaced by spruce and fir—another example of succession—and the climax or steady state peculiar to that altitude is again attained. In like manner, each of the fully developed vegetation zones in the mountains, in spite of temporary internal disturbances, are self-perpetuating communities of living things. In their climax form, no other combination of dominant species is ever successful in replacing a community adapted through time to a given altitude.

Mountains, with their varied exposures, differences in altitude, and climatic diversity, provide isolated habitats in which different species of plants and animals can exist. These living things also show diversity in form and activity since each must adapt to the environment where it lives. This involves its niche, which is its business of finding food, shelter, moisture, and mates in order to continue life. These activities in any given habitat result in competition between species and exchanges in energy and the development of food chains.

All food chains begin with photosynthesis, the process whereby green plants assimilate carbon dioxide into energy-rich carbon compounds in combination with mineral elements from the soil. Only plants can perform this function and consequently they are the *primary producers* in nature. An animal, be it an insect, mouse, or a deer, that derives its nutrition from plants is a *primary consumer*. A carnivore is a *secondary consumer* since it derives its energy from the plant eater or herbivore. Carnivores that feed on other carnivores are *tertiary consumers*—the eagle that catches the mink, that catches the muskrat, that eats only plants.

Energy is lost at each link in the food chain. A multitude of seeds are required to feed the mice on an acre of meadow. A few mice suffice to feed a weasel. And an occasional weasel is enough to feed an owl. The food chain, however, does not end with the owl, the eagle, the bear, or the fox. When these animals die their bodies are eaten by magpies, crows, and skunks, or their substance is returned to the earth by bacteria and fungi and their organic matter is converted to mineral form to be recycled through other plants and animals in the mountain ecosystems.

Plants and animals have been important components of the geological-biotic system for millions of years. The coal, peat, oil, and even the carbonate rocks we use today were generated in the past from plant and animal remains. Mollusc shells and the hard coverings of protozoans have contributed to limestone formations. Even some aquatic plants release calcium carbonate as a by-product of photosynthesis. Thus, some of the substance of mountains comes from producer organisms that lived in the past.

In the present, man is one of the main consumers of materials that come from mountains. The trees that grow there produce lumber for houses, ties for railroads, paper for industry, books, and even Christmas cards. Game animals, including deer, elk, bears, and rabbits, provide recreation and meat for the dinner table for millions of sportsmen. Streams that flow with clear water because of the filtering effect of vegetation furnish power for hydroelectric power plants, water for crop irrigation, and drink for millions of people in towns and cities. And not least of all, the very landscape of mountains and its interacting geological, biotic, and climatic components provide recreation and outdoor enjoyment for ever increasing numbers of human beings. All these products come from the interaction of geological and biological processes that are still in progress, although they began aeons ago.

It was the original convulsions and upheavals that started the formation of the earth's surface as we now know it. In the beginning the waters came out of the steamy mists and out of the rocks of the world. Above the crust of the planet, Pangaea—a super continent—appeared in the midst of Panthalassa, a universal sea. And two-fifths of the face of the earth was land and three-fifths was sea.

Then enormous upheavals beneath the earth's surface made cracks in the crust and separated Pangaea, the giant merged continent, into two continents, Laurasia and Gondwana. The first of these give birth to Asia, Europe, and North America. The second part included the land that was to become Antarctica, South America, Africa, and India. Then Gondwanaland separated and the part which became India moved to join Asia. The remaining part separated to become South America-Africa as one continent and Australia-Antarctica as another continent.

But the motive forces within the earth still were not done. Australia and Antarctica separated about 120 million years ago. The separation of the other super continents continued through geological time. In this colossal spreading apart of the land masses South America and Africa moved apart, and they are still separating at the rate of two centimeters a year. North America moved away from Europe, leaving between them the Atlantic Ocean, and they are now separating at a rate of 1.2 centimeters a year.

How do geologists know of these movements of the continents? And how do they explain this spreading of the seas? The evidence from paleomagnetism and geophysical study of oceanic basins, obtained by cores drilled from the depths of the seas, has revived the idea of sea-floor spreading and continental drift. The story, which is still unfolding, involves the belief that the earth's crust is divided into rigid plates of rock—the theory of plate tectonics—that float near the surface, crunch into one another or scrape sides, pushing up mountains and causing earthquakes. Some plates slide beneath one another and become molten as they disappear into the earth, finally to rise again from great faults and ridges in the floor of the sea.

The propelling force that moves the plates and causes continents to drift comes from mighty convection currents induced by radioactive decay and geothermal energy within the planet itself. The old geology told us our planet was a dying, shrinking globe, wrinkling its skin into mountains like a drying apple. The new geology tells us that land forms still are being uplifted, sculptured by erosion, deposited in the sea, and being thrust into the depths of the earth to be melted and born again as the tectonic plates wander over and under the surface of the earth. The lesson here is that mountains are not eternal. And man's association with them is no more than a one second tick in time.

If one looks at mountains with a broad view it becomes apparent that topography consists essentially of ridges and valleys. In mountain landscapes, ridges are higher and valleys are deeper than they are in plains and deserts. The overall pattern is accentuated in mountains where streams and ridges form easily discernible local drainage basins each with its own valley.

My geography teacher in college used to say, "Every valley is the work of the river it contains." There is an element of truth in this statement, but it tends to oversimplify the explanation of how landscapes have originated. The basin of the Rio Grande River in New Mexico, for example, was not sculptured by the river alone. Instead, it originated as an elongated depression of the earth's crust between two faults; geologists call such a trench a graben.

The continental glaciers moving down from Canada greatly influenced landscapes in the northern United States. They deposited moraines that changed the courses of rivers. They rounded off mountains and exposed rocks that were flat or tilted and folded. Flat layered rocks tend to be eroded into plains or plateaus; folded rocks, consisting of soft and hard layers, tend to be eroded into long ridges with the hardest rocks forming the highest ridges. To understand these land forms we must remember that among the keys to their features are geological history, climate, gravity, and especially water.

Water is one of nature's most important sculpturing agents. In streams and rivers it is a cutting tool with the power to wear away the hardest rocks. Fast moving water with its load of sediments grinds down boulders and transports rock particles toward the sea. Hell's Canyon, the deepest in North America, was created by rain and the running water of the Snake River between Oregon and Idaho. Through differential erosion water leaves most of the hard rocks of mountain peaks standing high above the horizon while it erodes the softer stones of the mountainsides and of the foothills. The pebbles, sand, and silt loosened from mountains are finally laid to rest in the sea where in future ages they may again change to rock and be uplifted into mountains again.

Wind is another of nature's forces that shapes rock formations. With sand as a grinding tool it erodes solid rock, shifts sand dunes, and redistributes soil over land and sea. Wind in the mountains sweeps rock particles off slopes and deposits them in valleys. It also carries glacial dust hundreds of miles and leaves it in deposits, called loess, which stand in vertical cliffs when streams cut through their depths. The loess of the Mississippi

The Colorado River in Glenwood Canyon has cut down through horizontal layers of Cambrian Sawatch sandstone which lies on Precambrian granite. These layers are ancient beach deposits of quartz sand, colored with green mineral, glauconite, and the red mineral, hematite. The sea covered this area late in the Cambrian Period.

Valley is believed to be the material of the Rocky Mountains, transported by water and wind during and following the Ice Age. In similar manner the ashes and fine grained materials of the western volcanoes were scattered far and wide over eastern Oregon and Washington by the prevailing winds from the Pacific Ocean.

Thus, the landscapes we now see were developed in the past. Mountains were uplifted millions of years ago. Most of the glaciers have disappeared. Volcanoes of the West are mostly dormant. But still the weathering processes continue, readily apparent as we explore the great outdoors. Sand and dust blow into our faces as the wind blasts through rocky canyons. A clump of sod drops from the edge of a gully when the rain beats down. The surface of smooth rocks flakes off in shell-like layers under the action of wetting, heat from the sun, and the frosts of winter. A rock falls from a cliff, adding to the talus or sloping deposit of rocks below. Marble and limestone dissolve in the presence of carbon dioxide dissolved in water. Even the rivers carry lime and minerals in solution as their clear water goes down the mountain and flows to the sea.

The disintegration of rocks through physical and chemical change provides the source and foundation of soils upon which all living creatures depend. From the biological viewpoint soil is a combination of weathered material of the earth's crust and organic material consisting of living organisms and the products of their decay. Soil not only contains living organisms but is produced by them. And soil influences the kind of vegetation that can grow in any given locality. The kind of plant cover in turn influences the appearance of the landscape.

Soils, like geological strata, consist of layers. They have a history of development which is related to time, topography, climate, source of parent material, and nature of the plant cover. The original material may consist of rocks disintegrated in place, or it may have been transported by gravity, water, glaciers, or wind. Whatever its origin, if it remains in place long enough it develops a profile of layers called soil horizons. Soil scientists have names and detailed descriptions of these layers.

Soil types and profiles are greatly influenced by the nature of the topography, kind of parent material, climate, rate of decay of organic materials, and drainage. Bogs develop where poor drainage and insufficient aeration slows down disintegration of organic matter and favors accumulation of humus. Steep slopes, where drainage is rapid, exhibit soils with thin horizons. Alpine tundra soils have shallow profiles with high percentages of gravel and undecomposed organic materials in their cold windswept habitats. Mountain landscapes thus become more meaningful when we understand that variations in soils and climates create innumerable environments with differences that influence the lives of plants and animals and thereby contribute to biological variety in the mountain world.

Much of the variation and beauty of mountains relates to their form and mode of origin. In general, there are four kinds of mountains—volcanic, dome, fold, and fault-block. Most of the volcanic mountains lie along the Pacific Coast and in the Southwest. Other kinds of mountains, such as the Rockies, are very complex and are combinations of folding, volcanic activity, and granitic uplift. Dome mountains, on the other hand, are relatively simple in structure and consist of a single upfold.

The Black Hills in South Dakota and Wyoming are dome mountains. These were formed by uplift of deep rock formations that pushed up and cracked the surrounding layers as the central core bulged from deep within the earth. Dome mountains are formed by pressure of molten rock deep in the earth. As the molten mass rises through a fracture it expands without coming through the surface layers. Consequently, it forms a bulge. If subsequent erosion removes the surface strata the hard rock of the dome may be exposed.

The turned up edges of the rock formations around the dome

The Ruby Range and other mountains in the Great Basin in Nevada rise above the sagebrush desert with essentially barren slopes. Forests, lakes, and meadows occur near the summits.

now form concentric rings of foothills that increase in height toward the center of the mountains. Some of these concentric rings of rock layers are hard and resistant to erosion. The softer, more easily eroded formations between the hard ridges become worn down so that river valleys tend to run in circles around the central dome. Other dome mountains are the Zuni Mountains in New Mexico, the Henry Mountains in Utah, and the Adirondacks in New York.

Fold mountains are nicely represented in the Valley and Ridge Province in the Appalachian Highlands. The Ouachita Mountains in eastern Oklahoma also were formed by folding. Fold mountains result when crustal plates beneath oceans and continents push together and wrinkle the earth's surface. When these wrinkles are numerous, as in the Appalachians, a series of parallel ridges with intervening valleys is formed. Erosion has left the resistant rocks so they form mountain ridges. These folded mountains may be hundreds of miles in length. A glance at the highway map of Pennsylvania shows how the folded Appalachian Mountains cause the roads to run northeast to southwest in narrow parallel valleys.

Many of the world's great mountain ranges have been formed in part by folding of horizontal rock strata, sometimes to great heights, as in the Alps, the Himalayas, and the Andes. The Rocky Mountains have been folded, uplifted vertically, and subjected to volcanism in some localities. The crustal movements of the Front Range of the Rockies west of Denver have been up and down, and during the last half billion years these mountains have been folded, eroded, encroached upon by the sea, and uplifted again. Here, the combination of mountain building forces —uplift, volcanism, erosion, deposition by streams, and sculpturing by ancient glaciers—has given us an admirable and fascinating geological story.

Block mountains are the common land forms in the Great Basin. Their structural pattern is the result of faulting and crustal movement. The typical block mountain has a steep face along the fault line and a long sloping side away from the faulted side. The sheer face of the mountain usually is the uplifted edge of the rock layers and the gently sloping side is the result of subsidence. Fault-block mountains provide some of the world's most spectacular scenery. The steep eastern faces of the Sierras in

California and the Tetons in Wyoming are two of our best known examples. They, like the Rocky Mountains, were formed by a combination of faulting, uplift, volcanism, and other mountain building forces.

Some of our most rugged scenery is found in the great plateau regions. Plateaus essentially are raised plains of high relief. They are uplifted by diastrophism, or movement of the earth's rock crust. Some plateaus result from vertical faulting which leaves the rock layers in more or less horizontal position. The Colorado Plateau, for example, is an enormous fault plateau covering thousands of square miles in western Colorado, Utah, and northern Arizona. River erosion has cut many valleys through this plateau, including the Grand Canyon. Because of the high relief and the rugged topography this plateau has many of the characteristics of a mountainous region.

The Appalachian Plateau in eastern United States is a mature plateau, marked by high rolling topography and by valleys rounded by millions of years of erosion. The mountains of this

The Maroon Bells with their tilted red sedimentary ridges are two of Colorado's most beautiful mountains. Glaciation has rounded the valley in the foreground. Maroon Peak is 14,159 feet and North Maroon Peak is 14,014 feet. Many mountaineers have lost their lives by falling from these peaks.

region are forested to their skylines as may be seen in the Cat-skill Mountains in New York, the Allegheny Mountains in Penn-sylvania, and the Cumberland Mountains in Tennessee and Ken-tucky.

The results of mountain building by volcanoes may be seen in many places in the United States. Mauna Loa, on the island of Hawaii, is the largest active volcano in the world. Kilauea, which rises 20,000 feet above the floor of the Pacific, is one of the most active. The Aleutian Islands and Alaska contain active volcanoes, some of which are the most beautiful in the world. The Cascade Mountains of Washington, Oregon, and California contain many well known cones, including Mount Baker in the north, Mount Shasta, and Lassen Peak in the south.

Ancient volcanoes occur in the Datil and Jemez Mountains in New Mexico, in the San Juan Mountains in southwestern Colo-rado, in the San Francisco Mountains in Arizona, and in the mountains of Nevada. Great lava flows are widespread in east-ern Washington and on the Snake River Plains of Idaho. The most famous area of still somewhat active volcanism is in Yel-lowstone National Park where geysers, steam vents, and hot springs attract thousands of visitors each year.

Few men have witnessed the birth of a volcano. But Dionisio Polido saw the first smoke of El Paricutin beneath his feet near the village of San Juan Parangaricutiro, Mexico in February 1943. Soon the earth trembled. From a crack in the ground came cinders and glowing rocks. Then the rumbling fuming mountain began to form. Its cinder cone grew to 1,000 feet within ten weeks and hot lava flowed from its side. In June, violent explo-sions tossed dust and rock bombs into the sky.

The terror of volcanoes in eruption has been known for mil-lennia. Legends of the Paiutes tell us how magic mountains poured fire and molten rock from Mount Shasta, Mount Ma-zama, and other volcanoes of the Cascades and the Sierra Ne-vada 10,000 years ago. The heat from these eruptions burned the skins of the Paiute ancestors dark brown. Mount Vesuvius buried the towns of Pompeii and Herculaneum beneath ashes in 79 A.D. when the Roman Empire was in its glory. One of the most violent explosions in all history occurred in 1883 when the island volcano of Krakatoa in the Indian Ocean disintegrated with a roar heard 3,000 miles away. The even larger volcanic exposion of Thera, an island in the Aegean Sea, occurred around 1,200 B.C. Its caldera is five times the size of that of Krakatoa. Archaeologists speculate that this explosion may have affected the Minoan civilization of ancient Crete and possibly may have inspired the Atlantis myth.

In more recent time, on May 8, 1902, Mount Pelee, on the French island of Martinique, destroyed the entire city of St. Pierre with its 28,000 inhabitants. In 1963, the volcanic island of Surtsey off the coast of Iceland raised its fiery crest above the sea. Ten years later, in 1973, Helgafell volcano covered Vestmannaeyjar, an Iceland fishing village, with ash and closed the harbor to the fishing boats.

The eruptions of fire mountains produce lethal gases, dust, ashes, rock bombs, and the irresistible molten lava flowing down the mountainside, covering all things in its path. Violent explosions release gas pressure in the rocks. This in turn releases other explosions that shatter magma into dust. Superheated steam and dust combine to make the black clouds that pour out of volcanoes. When the steam cools to rain the dust is wetted and falls as mud, accompanied by lightning produced by friction between rock particles in the air. The glowing red and yellow rocks and cinders that burst out of the fiery earth with the roar of cannons are awesome accompaniments to the sight of the very mantle of the earth rising into a mountain.

Several hundred volcanoes of the world are still active. Vesuvius, the one of ancient history, is still alive as is Mauna Loa in Hawaii. Mount Shasta in northern California is active below its 14,162 foot summit and even Mount Hood in Oregon has its hot spot that melts the winter snows. Other mighty volcanic mountains, including the Spanish Peaks in Colorado, Mount Jefferson in Oregon, and Mount Baker and Mount Rainier in Washington are silent, at least for the present.

These volcanoes are only a part of the great belt of fire mountains encircling the Pacific Ocean in a pattern that parallels the western coasts of South and North America, crosses the Aleutians to the Asian coast and continues on to Australia and New Zealand. The symbol of all Asian volcanoes, of course, is the perfect cone of Fuji in Japan. Each volcano, however, has its own characteristics, depending on its time and place of birth and its geologic history.

The ultimate cause of volcanoes is still conjectural; however, the recent theories of sea-floor spreading, continental drift, and plate structure of the earth's outer shell offer plausible explanations of how crustal movements produce upheavals that culminate in mountains and volcanoes. When continental-sized plates tear apart, or collide so that one thrusts beneath another, great faults may be produced through which hot magma may rise to the surface.

The mother material of volcanoes is magma, or basalt, which may be heated by high temperatures generated by radioactive decay of uranium and thorium. The heated rocks remain solid under extreme pressure miles below the surface of the earth. Their substance includes silica, alumina, magnesium, iron, calcium, sodium, potassium, and water vapor. When a rift occurs in the earth's surface the rock beneath becomes semi-fluid and rises from the depths through faults and fractures. The molten rock also may expand several thousand feet down to form a magma chamber or reservoir from which future volcanoes can erupt. But if a fracture extends deeply into the earth's crust the magma may come to the surface and flow over hundreds or thousands of square miles to form a lava plateau such as the one now covering much of eastern Washington.

Volcanic eruptions occur when pressure on heated rocks is reduced, allowing the contained gases to come out of solution, expand, and fracture the roof of the magma chamber. Then the magma pours to the surface. If the vent to the surface becomes plugged by solidified lava, pressure may continue to develop in the magma chamber until the volcano explodes. Pumice, ash, gas, clinkers, rock bombs, and blocks of lava then are blown into the air while fluid lava flows down the slopes in a fiery glowing avalanche.

Volcanoes occur in many shapes, depending on the materials they erupt. Basaltic lavas that flow rapidly from the vent spread widely over the landscape and produce broad rounded mountains such as Mauna Loa and Kilauea in Hawaii. When ash, clinkers of lava, and bombs are ejected they pile up into cinder cones with steep sides, usually having a conical depression at the summit. Volcanoes that erupt periodically frequently are composed of alternating layers of ash, pumice, siliceous fragmented ejecta, or lava.

If the rim of the circular depression at the summit of a large volcano collapses and falls into the vent, a caldera or depression is formed in the top of the mountain. The caldera of Crater Lake in Oregon contains the deepest lake in the United States. Within this lake is Wizard Island, a cinder cone produced by an eruption that occurred after the caldera was formed.

Each volcanic mountain has its own history of birth, development, and old age. Some volcanoes have complicated biographies. Mount Rainier which thrusts its white summit nearly three miles into the sky owes its spectacular scenery to many events. Some sixty million years ago there was no Cascade Mountain range nor any Mount Rainier. The lowlands of that region bordering the Pacific Ocean apparently were sinking and swamps with luxuriant vegetation were being covered with sand and clay brought in by rivers flowing from the east. Sedimentary beds of coal, shale, and sandstone accumulated to a depth of 10,000 feet.

About 40 million years ago volcanoes appeared on this coastal plain and then sank beneath the sea. Their outflows of lava produced rock layers nearly 10,000 feet thick which ultimately were uplifted, folded, and then eroded into deep valleys by many rivers. A few million years later, volcanic eruptions began again and covered the landscape with pumice to form a broad volcanic plain. Still more volcanism followed with eruptions of basalt and andesite, after which the volcanoes became extinct. Then once more the rocks were uplifted, folded, and fractured and then more molten rock pushed upward. Most of it hardened below ground where it became the foundation for Mount Rainier.

The birth of Mount Rainier began several hundred thousand years ago with huge lava flows and gigantic explosions that tossed large rocks for many miles. Flow after flow piled lava in layers and built the giant cone. The Ice Age came and went. Pumice still came out of the volcano 10,000 years ago. This region was also covered with pumice from Mount St. Helens in southern Washington and from Mount Mazama far to the south in Oregon. Mount Rainier erupted again, 6,600, 5,800, 2,500, and 2,000 years ago. The summit cone is the result of these final major eruptions. Minor eruptions occurred only 100 years ago. Who can say that the magnificent mountain will not erupt again? Geologists believe it will.

Three main classes of rocks, each formed in a different manner, make up the substance of mountains. These are the sedimentary, igneous, and metamorphic rocks. Each of these includes minerals formed from the basic elements and chemicals in the molten magma deep in the crust of the earth. The combination of these elements in rocks and the ways in which rocks are formed determine the variety in mountain landscapes. They also determine our patterns of life since the existence of plant, animal, and human life depends on the soil which comes from rocks and on the minerals we use in multitudinous ways.

Most of us are familiar with sedimentary rocks when they occur as limestone, sandstone, or shale. These are formed of many materials. Particles of sand washed down by rivers to the sea settle in layers and become cemented together by pressure and heat and thus are consolidated into rock. Fine particles of older rock, deposited as mud, ultimately are metamorphosed into shale. Limestone is formed from particles of lime deposited in the shells of clams and other sea creatures. Sometimes, fossils of crablike trilobites and crinoid (sea lily) stems are preserved intact in limestone layers. When these fossils are found on mountain tops and canyon walls they tell us that the limestone layers were originally built beneath the sea.

Sedimentary rocks consisting of rounded pebbles and gravel cemented together by minerals such as silica are called conglomerates. The cementing material is brought in by water percolating between the pebbles of quartz, granite, or other minerals. Some conglomerates are soft enough to be crumbled easily by hand or with a hammer. Others are hard and firmly bound together and are resistant to erosion.

Most sedimentary rocks have been deposited in nearly horizontal layers. Where deposition continued over millions of years, as in western Wyoming, the rocks subsided below sea level and as much as 30,000 feet of sediments accumulated. In geologic time most sedimentary rocks have been tilted, frac-

Many limestone layers are exposed in Vermont. Water dissolves lime from cracks and crevices and frost action aids in the disintegration of these stone walls.

tured, and in some instances subjected to great pressure. Through these processes limestone in some localities has been metamorphosed into marble and shale into slate. In addition, plants that grew in ancient swamps, and were compacted and altered to coal, are properly called sedimentary rocks. Coal under pressure from folding, as in western Pennsylvania, has formed anthracite, the hard long-burning coal our grandfathers used in their base burner stoves in their living rooms. Now coal is once more becoming a major energy source in America.

The sedimentary rocks of desert mountains provide opportunities for study of prehistory of the earth. Imbedded in these rocks are many fossils, including bones of dinosaurs, horses, camels, elephants, and sea shells. Erosion even uncovers "fossilized" sand dunes with their strata undisturbed although they have been buried beneath other sediments for millions of years.

Igneous rocks, formed from molten magma deep within the earth, give us some of our most fantastic mountain scenery. The Yellowstone Park area, with its geysers, cliffs of volcanic glass, and twisted columns of lava, exhibits hundreds of square miles of igneous materials ejected in some of the greatest explosive eruptions in all geological time. The giant fluted columns of rock that form the Palisades across the Hudson River from New York are igneous rocks. The Spanish Peaks, also called *Waha-toyas* or woman's breasts, in southern Colorado rise to 12,708 and 13,623 feet respectively. These are the solidified cores of ancient volcanoes that may have been the tallest ever developed on the continent. The emerging lava tore out and filled radial cracks in the earth. These now are exposed as dikes in the "wall rock country" near Walsenburg. Other great volcanoes composed of igneous rocks are the great cones of Mount Baker and Mount Rainier in Washington, Mount Hood in Oregon, and Mount Shasta in California.

Igneous rocks are of many kinds, depending on temperature, pressure, and the composition of the magma at the time of their origin. Magmas generally contain much silica and varying proportions of elements such as iron, magnesium, aluminum, calcium, sodium, and potassium. Igneous rocks low in silica generally are dark and form basalt. The lava which covers much of the Yellowstone Park area contains much silica and is light in color. As it comes from the earth it is a taffylike almost liquid substance which solidifies into ridges and concentric patterns. Magma that cools rapidly above the surface may form glass or obsidian without crystalline structure. Granite, on the other hand, is crystalline and commonly is the result of intrusion of melted rocks into fissures in the earth's crust. Some granites, however, are formed by metamorphism.

Metamorphic rocks are formed from other rocks by combinations of heat, pressure, and chemicals. These new kinds of rocks may be produced either from sedimentary or from igneous rocks. Granite, for example, may be metamorphosed into gneiss, a rock in which the minerals are arranged in layers. Shale may become slate. Limestone may become marble. And basalt may become greenstone, a rock colored green by chlorite, hornblende, and other minerals. The great heat and pressure that produces metamorphic rocks can also recrystallize the contained minerals into new kinds of minerals.

Some metamorphic rocks contain numerous kinds of minerals. Gneiss, for example, is a mixture of feldspar, mica, hornblende, and quartz. Schist, with its high mica content, is a platy rock and sometimes can be cleaved with a pocket knife. If schist is exposed to great heat and pressure it, in turn, can be melted to magma and metamorphosed to granite. Thus we can understand how the creaking, groaning, erupting, eroding earth has produced and arranged its rocks and mountains in all their variety and beauty.

The list of rocks and minerals, including their crystalline forms, appears to be endless, especially if you attempt to assemble a representative collection of them. Soon they begin to fill your basement, and possibly your garage. Then as you at-

Fluted columns of igneous rock form Devils Tower in northeastern Wyoming. This strange formation rises more than 1,000 feet above a base of sedimentary rocks. Cracks and joints were formed when the rock columns cooled.

tempt to name and classify them you become aware of their multitudinous physical and chemical variations and of how they were formed as the mountains themselves came into being. Rock hounds and gem stone collectors are especially adept at finding choice stones and minerals since they are knowledgeable about the forces that have wrought the materials of mountains. You do not have to be a geologist or a mineral expert, however, to appreciate the diversity, complexity, and beauty of the basic materials of the earth.

A piece of granite offers an excellent opportunity for examination of common minerals and for contemplation of the processes and forces that brought about its formation. You can see granite monuments in almost any cemetery in the nation. Many of the state capitol buildings are built of stone and contain granits. You can see granite in innumerable places: in the granite quarries in Vermont; on Stone Mountain in Georgia; in the walls of Big Thompson Canyon on the way to Estes Park, Colorado; in the Teton Mountains in Wyoming, in the Sierra Nevada; and in hundreds of other mountain ranges from the Atlantic to the Pacific coasts. Wherever you find granite, with all its variations in texture and color, its predominant minerals will be quartz, feldspar, mica, and hornblende.

Variations in the proportions of these minerals result in red, gray, or speckled granites. The shining surface of polished granite is produced by the glassy quartz grains which form the background mass of the hard resistant rock. Feldspar crystals themselves have varying chemical compositions and their shiny surfaces are gray, red, or even white. They also vary in size—some are more than an inch in length. Thin plates of mica range in color from transparent to gray or brown. Hornblende, which contains aluminum and is found in basic igneous rocks, commonly occurs as green or black glassy crystals.

Gypsum is a white mineral which occurs as crystals of calcium sulfate. Calcite, consisting of calcium carbonate, is colored yellow, green, or orange by various impurities. Other nonmetallic minerals found in rocks include sulfur, graphite, fluorite, asbestos, and halite which contains potassium, bromine, magnesium, and other salts along with the ordinary salt we use in food.

The metallic minerals include a great variety of metals of which many are indispensable in industry. Copper, iron, gold,

Mesa Verde and Mancos shale formations form the slopes below resistant sandstone layers near Grand Junction, Colorado. This type of weathering and erosion is characteristic in arid climates.

and silver have been mined since ancient times. Other valuable minerals are tin, nickel, chromium, zinc, uranium, and aluminum. Some of these occur as veins, flecks, plates, or pure nodules in rocks. Others form complex chemical compounds with various elements. Aluminum assumes various mineral forms in combination with oxygen, sodium, and water. Bauxite, the common aluminum ore found in Arkansas, is a group of white oxides which sometimes are stained red by iron oxides.

Some of the most extensive mineral deposits are known as mineral fuels. Coal, for example, is abundant in Pennsylvania, Colorado, Wyoming, and New Mexico. Now that the nation is in an energy crisis the oil shales of western Colorado, southwestern Wyoming, and northeastern Utah will undoubtedly be exploited as a source of crude oil. These sedimentary deposits have been estimated to contain fifty times as much oil as has already been produced by drilling or held in reserve in the United States.

The oil in this shale has a strong odor that can be detected even in a rock picked up from the ground. There is a story, possibly apocryphal, of a man who built the fireplace in his house with shale rocks. When the first fire was started in the fireplace, the stones literally burned and the house was consumed with them. We can believe this is possible when we learn that a ton of processed shale may yield nearly a barrel of petroleum and several thousand cubic feet of natural gas.

The formation of minerals has long been a puzzle both for geologists and for chemists. The story of metamorphism, whereby minerals change into other minerals, involves not only the titanic forces that readjust the crust of the earth but the very atoms of which all things are made. Forces almost beyond comprehension cause magma intrusions between layers of rocks. When different layers remain in contact, wandering atoms congregate and form particles or masses which become minerals. Other substances of the earth also go through many changes. Clay, for example, hardens into shale. Under continued pressure shale becomes slate, then schist, and then gneiss which is a crystalline rock.

In magma that cools slowly, crystals have time to grow. But lava from volcanoes sometimes cools so rapidly that glassy obsidian instead of crystals is formed. Each volcano produces a slightly different kind of obsidian, a fact that has enabled archaeologists to trace trade routes in the Mediterranean region. Obsidian arrowheads also are found far from the original sources, indicating that Indians once traveled widely in the American Southwest.

The shape and other characteristics of crystals depends largely on the arrangement of atoms within the molecules of the crystals. Since some 2,000 minerals, many of them crystalline in structure, have been identified, and since many are beautiful and some are rare, it is not surprising that thousands of people make a hobby of collecting, cutting, and polishing them.

Experienced gemstone collectors are well aware of the characteristics of interesting specimens. Among important physical properties of gems are color, crystal form, hardness, cleavage, luster, specific gravity, index of refraction, and chemical composition. Color is one of the first attributes noticed in gemstones. "Ruby" was once applied to all red stones; "sapphire" to blue stones; and "topaz" to yellow stones. Names of gemstones now are based on color, luster, and other properties of the gems. Some are even named for persons.

Luster refers to the way in which the mineral reflects light. Hardness and heaviness are also diagnostic characteristics: iron

Gemstone hunters seldom find quartz crystals as large as this one taken from a mine in the Ozarks in Arkansas. The mountains are great hunting grounds for enthusiastic "rock hounds" who combine a knowledge of geology with enjoyable outdoors exploration and exercise.

ores are heavy; calcite is soft and can be scratched with a thumb nail; diamond is the hardest substance known. It scratches all other minerals. Fracture and cleavage define the ways in which minerals break. Clay crumbles; obsidian shows conchoidal fracture, leaving surfaces that resemble clam shells; galena cleaves into rhombohedral forms, which you will recognize if you remember your geometry.

With the boom in "rock hounding" these days, thousands of people are taking up the hobby of collecting minerals and searching for gemstones. Some of the best hunting grounds are remote and seldom visited. One of my favorites is central and southeastern Oregon. The Jordan Craters in Malheur County have spectacular displays of recent basalt lava flows. Among the interesting features are cinder cones, pumices, and lava bombs. The rim rocks are made of reddish rhyolite and if you know where to look you can find thunder eggs, agates, and obsidians.

If you live near mountains or go to mountains for vacation trips, time spent in exploring for rocks and minerals will reward

The Book Cliffs in eastern Utah form an escarpment about 2,000 feet high and 100 miles long at the southern edge of the Uinta Basin. The Cretaceous rocks of these cliffs dip downward under Tertiary formations, which are exposed to form the escarpment known as the Roan Cliffs farther north. In the desert foreground, the author examines a harvester ant mound.

you with a better understanding of the materials and processes that make our earth a remarkable planet. Wherever bedrocks are exposed, cliffs rise out of valleys, rocks are tumbled in rivers, or lavas radiate from ancient volcanoes, you will find history of the earth and the substances of which it is composed.

Mesas are flat topped mountains, bounded at least on one side by a steep cliff and capped by layers of erosion resistant rock.

2

America's Mountains

THE MOUNTAINS OF AMERICA have many faces. Differences in geology, age, and origin of the rocks, and climate give our highlands an enchanting diversity of scenery and life forms. The tree clothed rounded peaks of the Adirondacks, with their thin mantle of rock and innumerable lakes are geologically related to the great Laurentian Upland that extends northwestward to the Arctic Ocean. The Appalachian Highlands are a complex of old and mature mountains, with young rivers left by glaciation, imposing monadnocks rising above ancient peneplains and layers of stone folded into a seemingly endless series of long parallel ridges and valleys. From Mount Katahdin in Maine, the White Mountains of New Hampshire, the Green Mountains of Vermont to the Great Smoky Mountains and Mount Mitchell in North Carolina the Appalachians display a fascinating variety of land forms, plant and animal life, and human culture dating back to Revolutionary times.

We ordinarily do not think of mountains as being part of the heartland of America. Half of the United States consists of plains, the midlands of the country. Yet within the plains the domed Black Hills thrust up above the surrounding country with their upturned fringes of sandstone surrounding their gold- and silver-bearing igneous cores, with their forests of oak, aspen, and western conifers, and with their harsh human history. Lesser uplifts such as the Turtle Mountains in North Dakota give variety to the landscape, along with buttes, mesas, and badlands which display fantastic scenery, even though they are not true mountains. The Ouachita Mountains of Arkansas and eastern Oklahoma, however, rise to respectable heights and are geologically similar to the folded Appalachians far to the east. North of the Arkansas River lie the Ozark Plateaus, deeply dissected by river valleys, lavish in their mixture of northern and southern trees, and exciting because of their wilderness character and interesting mountain people.

Standing as a mighty wall at the western edge of the Great Plains are the magnificent Rocky Mountains. In comparison with the Appalachians the Rockies are young mountains, characterized by complexity of structure, alpine topography, extensive glaciation in the north, and scenic appeal to adventurers of every occupation and age. Mining, lumbering, grazing, hunting, skiing, and the tourist business are principal assets.

Some of the most distinctive parts of the Rockies are the Lewis Range in Montana (including Glacier National Park), the Grand Tetons of Wyoming, the Wasatch Mountains east of Salt Lake City, Utah, and the Front Range in Colorado (including Longs Peak in Rocky Mountain National Park) and Pikes Peak west of Colorado Springs.

West of the Rockies lie the great plateaus, some of which are loftier than the Appalachians. The sedimentary strata of the Colorado Plateau, covering western Colorado, western New Mexico, southern Utah, and northern Arizona, lie 4,000 feet thick on top of 2,000 more feet of granite. The plateau is mountainous in that volcanoes have lifted their cones above the genneral surface; other mountains have been domed by uplift, and spectacular canyons have been carved by rivers. The most famous, of course, is the Grand Canyon of the Colorado River. The mile thick lava of the Columbia Plateau, covering southern Idaho, eastern Washington, and eastern Oregon, also has been incised by the Snake and Columbia Rivers to form mighty gorges such as Hells Canyon which is deeper than the Grand Canyon.

South of the Columbia Plateau, the Great Basin is bounded on the east by the Wasatch Range in Utah and the Colorado Plateau, and on the west by the Sierra Nevada and the Coast Ranges in southern California. Within this dessicated land of alkali flats and sagebrush deserts are more than one hundred ranges of fault-block mountains. Many of these impressively high north-south ranges are mantled with pine and spruce-fir forests below their summits.

Two great mountain chains form the Pacific Mountain system. From Canada to central California the Cascade and Sierra Nevada mountains form one of the most distinctive uplands on the continent. Uplifted in a gigantic fault-block in the south, marked by volcanic peaks of great heights and extensively glaciated valleys from south to north, these mountains are among the most beautiful in the world. Their slopes support some of the oldest and largest of known trees. And their canyons include the famous Yosemite Valley and the Tuolumme River Canyon. From the High Sierras, Mount Whitney (14,494 feet), lifts its peak higher than any other in the conterminous United States.

The other mountain chain, the Coast Ranges, rises abruptly from the Pacific Ocean and extends from the Olympic Mountains in northwestern Washington southward through Oregon and California. From north to south these ranges change from the wettest climate in North America to near desert conditions south of Los Angeles. Douglas firs grow into giant trees under the heavy rainfall in Washington and Oregon. Redwood trees flourish in the coastal fog of California.

The preceding summary gives only a generalized view of the principal mountain systems of the United States. The physiographers have systematized the designation of land forms by dividing the nation into physiographic divisions and these divisions into provinces. Some of these divisions do not concern us here, for example, the Atlantic Plain Division which includes the Atlantic and Gulf Coastal Plains, or the Interior Plains Division which includes the prairie and the plains country that forms the midlands of the nation between the Appalachian Plateaus and the Rocky Mountains.

You need not be a scientist to appreciate the geological, climatic, and biological diversity of mountains. But a general understanding of their origin and development can add to your appreciation of their major relief features, their beauty, and their majesty. In addition, if you wish to understand the relations between rock formations and their relative ages you should familiarize yourself with the geologic time table. This is based on

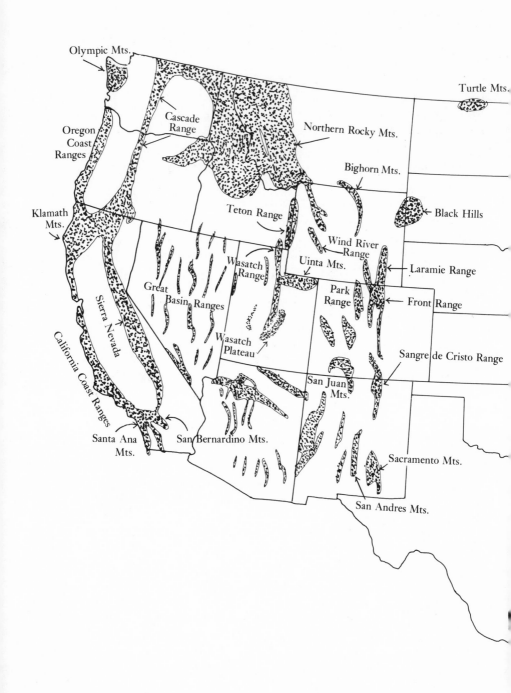

Olympic Mts.

Turtle Mts.

Cascade Range

Oregon Coast Ranges

Northern Rocky Mts.

Bighorn Mts.

Klamath Mts.

Teton Range

Black Hills

Wind River Range

Wasatch Range

Uinta Mts.

Laramie Range

Great Basin Ranges

Park Range

Front Range

Sierra Nevada

Wasatch Plateau

Sangre de Cristo Range

California Coast Ranges

San Juan Mts.

Santa Ana Mts.

San Bernardino Mts.

Sacramento Mts.

San Andres Mts.

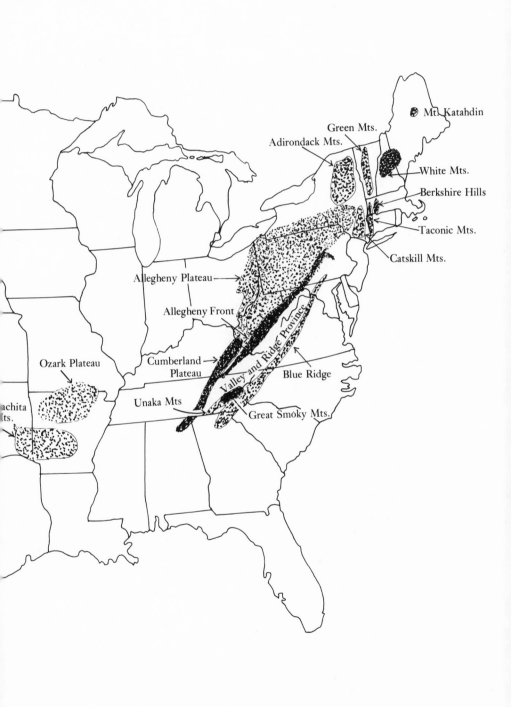

Mt. Katahdin

Green Mts.

Adirondack Mts.

White Mts.

Berkshire Hills

Taconic Mts.

Catskill Mts.

Allegheny Plateau

Allegheny Front

Ozark Plateau

Cumberland Plateau

Valley and Ridge Province

Blue Ridge

achita Mts.

Unaka Mts

Great Smoky Mts.

the sequence of rock layers, the bottommost layers being the first deposited and hence the oldest, and the topmost layers being the most recent and youngest.

The time scales used by geologists to date rock layers are based on the estimated time required to lay down deposits of sand, gravel, lava flows, beds of cinders, and other materials. Rocks transformed by heat, pressure, and bending of the earth's crust contort the original uniform and essentially level strata and make the dating process more difficult. An atomic time scale also is used now to estimate the age of layers in the earth's crust. This is based on the relative rates of decay of radioactive isotopes found in minerals in crystalline rocks. This new dating method is resulting in revisions of the time scale as study of earth history continues.

The generalized time table of geological history has been likened to the four seasons and twelve months of the year. Geologic time is divided into four eras: Precambrian, beginning some 4.5 billion years ago; Paleozoic, when fishes, amphibians, and land plants appeared; Mesozoic, the age of reptiles; and Cenozoic, the age of mammals, including man. Time since the Precambrian is divided into twelve periods. The periods are further divided into epochs, of which those in the Cenozoic Era are of special interest since they relate to the most recent history of the earth (Table 1).

The Adirondacks have always impressed me as being the gentle mountains. They offer one of the finest outdoor recreation areas in the nation. Their numerous rounded peaks, some of which rise to more than a mile above sea level, are majestic in their own setting but are easily climbed by ardent hikers. A fine network of hiking trails leads to countless lakes, carved by former glaciers, and to mountain tops from which other rolling hills and mountain peaks may be viewed in all directions.

On the way to the summits you encounter dense forests, pastured areas, and sparkling streams. Beaver ponds are present and if you are lucky you may see deer, raccoons, otters, and even a pine marten. On the trail to Mount Marcy (5,344 feet), the highest peak in New York, you may hear the charming song of the winter wren. Other birds to look for are the brown creeper, the red-breasted nuthatch, and the black-throated green warbler with its melancholy song. As you climb above the deep wooded valleys the forest becomes more stunted and alpine. Here the boreal chickadee and Bicknell's thrush appear. At the summit you are in an arctic-alpine environment where the timberline trees, balsam firs and red spruces, lie nearly prostrate with branches that grow like low spreading junipers.

Table 1. Geologic Time[1]

Time Divisions	Date of Beginning (millions of years ago)
Cenozoic Era	
Quaternary Period	2-3
Holocene Epoch	
Pleistocene Epoch	
Tertiary Period	
Pliocene Epoch	12
Miocene Epoch	26
Oligocene Epoch	37-38
Eocene Epoch	53-54
Paleocene Epoch	65
Mesozoic Era	
Cretaceous Period	136
Jurassic Period	190-195
Triassic Period	225
Paleozoic Era	
Permean Period	280
Pennsylvanian Period	320
Mississippian Period	345
Devonian Period	395
Silurian Period	430-440
Ordovician Period	500
Cambrian Period	570
Precambrian Era	4,500 +

[1] Based on *Geologic Time*, U.S. Department of the Interior, Geological Survey. Inf-70-1 (R4) U.S. Government Printing Office, 1972.

The many attractive peaks of the Adirondacks invite explorers because of the remote valleys and hills far from the miasma of civilization, the innumerable habitats for observing birds, the brightly colored flower gardens in late spring, the riotous tree colors in autumn. There is a group called the "Forty-Sixers" who rightly and boastfully claim they have ascended all forty-six Adirondack mountains 4,000 feet or over in elevation. The first three, Robert and George Marshall, and Herbert Clark, became Forty-Sixers on June 10, 1925. Since then approximately 600 have been certified and have had their names added to the roster of climbers. Their boundless enthusiasm, and their comradeship surely are symbolic of the spell of mountains, wherever they are.

Geologically, the Adirondacks are not a part of Appalachia, although the Lower Paleozoic formations around the granitic dome of the mountains is related structurally to the Appalachian Highlands. The core rocks of the Adirondacks are a southeastern extension of the Laurentian Upland which mantles much of Canada from the Labrador Peninsula to Manitoba, Saskatchewan, and parts of the Northwest Territories. The heart of the Adirondacks is a nearly circular dome more than one hundred miles in diameter, composed of Precambrian rocks, the oldest on the continent. This dome, which touches the St. Lawrence River and the western shore of Lake Champlain, comprises much of northern New York State.

The Adirondacks underwent successive uplifts while the sedimentary formations around them were being deposited during Paleozoic times. Glaciation subsequently rounded many of the valleys, leaving glacial moraines and many beautiful lakes. Blue Mountain Lake, for example, is considered to be one of the most beautiful lakes in the Adirondacks. Racquette Lake, noted for its bass, lake trout, and whitefish, is the largest, with a shore line of 99 miles. Other lakes of natural beauty are Long Lake, Lake Algonquin, and Lake Pleasant. Bear and deer hunting along the Adirondack Trail are said to be the best in the North Woods.

For naturalists and mountain climbers, wildlife, waterfalls, ice caves and panoramic views are there in abundance. The many stream valleys radiating from the Adirondack plateau support a varied vegetation from the high levels to the lowlands with a corresponding diversity of birds and mammals. For the casual traveler there are the peaceful villages in the valleys, the crystalline rocks with their bands of colored minerals, and the foothills forested with beech, white pine, paper birch, and sugar maple.

The White Mountains in New Hampshire are forested almost to the summit of Mount Washington and the peaks in the Presidential Range.

Appalachia, the mountainous part of the eastern United States, is a big country and an old country. The mountain-building processes that formed these complex ranges, subdued by aeons of erosion and now mantled with luxuriant vegetation, began beneath the sea in the Paleozoic era. As sediments from ancient mountains to the east accumulated, the sea floor sank and some eight miles of rock-forming materials were deposited. Ultimately the sedimentary formations were uplifted, folded, metamorphosed, and altered in many places by volcanic action. From Paleozoic time to the present the sea never returned and the high Appalachian mountains were planed down, uplifted again, and eroded for 200 million years, shedding their material to the Coastal Plain and rounding their peaks to the gentle profiles we see today. These picturesque and beloved mountains extend from Newfoundland to Georgia, a distance of some 3,000 miles.

The physiographers divide the Appalachian Highlands into provinces based on their rock structure, their land forms, and their geological origins. The Piedmont Province, lying at low elevations between the Fall Line and the foothills of the mountains to the west, stretches from southern New York to Alabama. Essentially it is a low plateau with many varieties of sedimentary, volcanic, and metamorphosed rocks below the land surface. Rising above the Piedmont Province to the west, the Blue Ridge Province includes the first ridges of the Appalachian Highlands. Its highest mountain is Mount Mitchell (6,684 feet), in North Carolina.

The Valley and Ridge Province, extending from the Hudson River in New York to the highlands in Alabama exhibits a fabulous array of folded and faulted mountains composed of rocks that vary from shale, sandstone, and limestone to anthracite coal. Beyond the Valley and Ridge Province and extending from the eastern front of the Catskill Mountains in New York, through northern Pennsylvania and eastern Ohio, and southward into Kentucky, Tennessee and Alabama, the Appalachian Plateaus exhibit folds that form the ridges and valleys to the east. The plateaus are dissected so deeply they are mountainous. The northernmost Appalachians lie in the New England Province which includes mountains of many kinds, varying from the Pre-

Fir forests cover the summits of the Great Smoky Mountains in the southern Appalachians. This view is from Klingman's Dome, North Carolina-Tennessee. Elevation is 6,643 feet.

cambrian cores of the Green Mountains in Vermont to the monadnocks of New Hampshire and the rocky valleys on the coast of Maine.

Upheavals of the ancient Appalachians hundreds of million years ago, differed from north to south, thus contributing to the [...] different ranges today. New [...] lome shaped granite uplifts [...] and ice action. The Alle- [...] ans now are parallel ridges [...] d crinkling of the earth's [...] r sculptured the mountains [...] rallel valleys and then cut [...] ow used by modern high- [...] e northern mountains also [...] s. Among the notable ones [...] Notch in the White Moun-

[...] are a rugged and varied [...] range is Maine's most out- [...] (5,268 feet) rises abruptly [...] r strewn country. Its sum- [...] the Oxford Hills which lie [...] To the north the Mahoosuc [...] Range include some of the [...] ing to absence of developed

[...] Hampshire provide some of [...] mountain scenery. Mount [...] timberline and is the high- [...] orth of the Carolinas. It is [...] violent weather, high wind [...] to the summit. From Pink- ham Notch Highway the road winds steeply through pictur- esque forests, past fold rocks and scrub timber, and emerges on the shoulder of the mountain from which a fine view of the Presidential Range includes Mounts Jefferson, Adams, Monroe, Madison, Clay and Jackson. All of these exceed 5,000 feet in ele- vation. The Mount Washington Railroad, completed in 1869, has one of the steepest grades in the world. Steam locomotives

An early autumn snow covers the rocks at the summit of Mount Washington in New Hampshire. The alpine summits of the peaks in the Presidential Range rise above the steep forested slopes in the middle background.

climb the three-mile cog railway over trestles and through scenic views in slightly over one hour.

Among the striking features of the Presidential Range, along with the Franconia Range to the west, are the U-shaped valleys and basins formed by glacial action during the Ice Age. Continental glaciers rode over the mountains and local glaciers continued their erosion after the last major advance of the ice. The notches cut by those glaciers add to the scenic beauty and provide access over paved highways.

Excellent views are to be had by driving through Dixville Notch in northern New Hampshire, Crawford Notch, and Franconia Notch. A fine view of this famous mountain pass may be had from Artist's Bluff above Echo Lake. Looking south in this rugged setting, Mount Lafayette lies to the east and Cannon Mountain lies to the west. Many hiking trails, described in the Appalachian Mountain Clubs' *White Mountain Guide*, diverge from these notches, affording climbers and explorers unexcelled opportunities for outdoor pleasure and healthful exercise in these historic and beautiful mountains.

Parallel to the White Mountains the Green Mountains occupy much of Vermont. This range is relatively steep but elevations are not as high as other Appalachians. Well known peaks are Mount Mansfield (4,393 feet), Killington Peak (4,241 feet), and Mount Ellen (4,135 feet). Geologically the Green Mountains are old and worn down; the ridge tops are rounded and forested with yellow birch, beech, maple, and oak. Scenic points of interest include Wallingford Waterfalls, the cliff of Mount Horrid, and the limestone escarpment known as White Rocks. Beneath the surface, gneissic rocks of Precambrian age form the core of the Green Mountains. These rocks also occur in the Hoosac Mountains in western Massachusetts, and in the highlands along the Hudson River.

The Green Mountains have many historical and recreational features. The first white man to visit Vermont was Samuel de Champlain who came up the Richeleau River from the St. Lawrence in 1609 and discovered the lake that bears his name. Ethan Allen, in 1770, organized the Green Mountain Boys to repel surveyors from New York who were disputing claims of settlers who had received land grants from New Hampshire. In 1775, Allen and his Green Mountain Boys surprised the British commander of Fort Ticonderoga in his nightshirt and captured the fort.

After the Revolutionary War thousands of families settled in the valleys, cleared land for farming, and logged the virgin timber on the slopes. New growth now covers much of the land on farms abandoned before the turn of the 19th century. Since the establishment of the Green Mountain National Forest the management expects to build up the timber supply and develop recreational use which now is the number one source of income in the state of Vermont.

An extension of the Green Mountains into western Massachusetts, called the Berkshire Hills, is a mixture of low mountains and extensive river valleys. The western part of the Berkshires lying along the Hudson River and Lake Champlain Valley is represented by the Taconic Mountains. Mount Greylock (3,491 feet) is the highest. Other interesting mountains, easily accessible by foot trails, are Mount Fitch (3,110 feet), Mount Sugarloaf (2,030 ft.), Brodie Mountain (2,613 ft.), and Rounds Mountain (2,198 ft.).

The eastern part of the Berkshires is called the Hoosac Mountain Range. The region essentially is a plateau with rolling topography and low mountains. The Hoosac range also traverses Massachusetts from Vermont to Connecticut. Both here and in the Taconic range to the west one can find much natural forest beauty, native culture, and historic heritage. In this easily accessible collection of mountains, points of interest include the Taconic and Mohawk Trails, Monument Mountain near Great Barrington, Pittsfield State Forest, Wahconah Falls near Dalton, High Falls in Savoy State Forest, and the Susan B. Anthony House at Adams.

In Connecticut, south of the Taconics, a group of small peaks, with pleasant streams and superb hardwoods, give one a feeling of the special beauty and tranquility of the countryside. These are the Litchfield Hills. Mohawk Mountain is here and from its summit on a clear day you can see the Berkshires in Massachusetts and the Catskills in New York. The Housatonic Valley also is here, bordered by lesser peaks such as Cobble Mountain (1,380 ft.). Throughout the wooded parts of this hill country one may expect to see deer, grouse, gorges jungled with stately hemlock, beech, and maple trees, and a great variety of ferns, grape vine thickets, and a variety of wild flowers.

South of New England the structure of the Appalachians changes. In the central part, fold mountains characterize the notable Valley and Ridge Province. In the Piedmont Province to the east the rocks are folded and faulted, but the low plateau is not mountainous. West of the high folded mountains, the Appalachian Plateaus are underlain with nearly horizontal rocks. The eastern boundary of these plateaus is formed by the Allegheny Front, an escarpment, or linear abrupt change in slope, overlooking the Valley and Ridge Province. The escarpment extends from Alabama to the eastern front of the Catskill Mountains in New York. This topographic break produces some of the most striking scenery in the Appalachians.

The traveler from the eastern seaboard can find Appalachian mountains close at hand. South of the Catskills in New York the topography includes highlands and great valleys. In New Jersey the uplands are not as imposing as those in Pennsylvania but they include some interesting peaks and the historic Delaware Water Gap near Columbia. Of interest are numerous peaks: Kittatinny, Hamburg, Copperas, Sparta, Scott's, and Jenny Jump Mountain. The Kittatinny Valley begins here in New Jersey and continues through Pennsylvania as the Lehigh Valley, the Lebanon Valley, and finally as the Cumberland Valley. Beginning in the Colonial period these valleys served as migration routes and places for settlement by thousands of immigrants to the New World. Above these fertile lands are the mountains of the Alleghenies extending westward toward Ohio, eastward into Maryland, and southward through Virginia and West Virginia.

The great mountain chain of Appalachia, with some of the finest mountains in the East, is the Blue Ridge which continues from Pennsylvania to South Carolina, Georgia, and Alabama. It is high country with ridges and peaks that rise above 6,000 feet. In the north, Shenandoah National Park preserves the biological heritage of the wilderness and the Skyline Drive along the divide provides views of majestic scenery for more than 100 miles. The rocks of the Blue Ridge are hard and resistant to erosion. Granite and quartzite, along with igneous rocks formed by ancient volcanic activity, define the topography.

The Blue Ridge separates into two chains south of Virginia. While the main ridge continues southward through the Carolinas into northeastern Georgia, the western prong becomes the Unaka Mountains in western Virginia, western North Carolina, and eastern Tennessee. In the southern part is the Great Smoky Mountains National Park, a Mecca for tourists and a biological paradise for naturalists. In this relatively unspoiled wilderness are hundreds of tree species, thousands of lesser plants, and a variety of animals ranging from salamanders to bears.

The oldest rocks of the Great Smoky Mountains, formed from sediments in a shallow sea, are so ancient they contain no fossils of plants or animals. These rocks—of a group called the Ocoee Series—are among the oldest on the continent and the uplift which built them into mountains occurred long before the Rocky Mountains were formed. The topography in the Great Smoky Mountains is the result of numerous processes. One of these is evident in Cades Cove where the Ocoee rocks were thrust over younger limestones for several miles. Subsequent erosion cut through the sterile ancient formations to reveal the fossils of primitive sea animals in the limestones of Ordovician age.

The whole land was uplifted several times and cut down to featureless plains. These flat land surfaces, created by long-continued erosion, are called peneplains. The ridges we now see in the Great Smokies are the remains of an uplifted peneplain and the valleys occupy the places where stream erosion has carved out the rocks of the high land surface. Rising high above the valleys are many well known peaks such as Mount Le Conte (6,593 ft.), Clingmans Dome (6,643 ft.), and other landmarks such as the Chimney Tops which may be seen from the trans-mountain road that crosses the Tennessee-North Carolina border.

There are other Appalachian mountains to be discovered and explored by the outdoor enthusiast. Some of these are disconnected from the main ranges and yet have an immensity all their own. Mount Mitchell (6,684 ft.), highest summit east of the Mississippi, rises from the Black Mountains in North Carolina. Georgia and Alabama also have mountains which are extensions of the Appalachians. Their ridges and peaks are rugged but mostly under 3,000 feet in elevation, although Springer Mountain, the southern terminus of the Appalachian Trail, reaches 3,820 feet. Steep and beautiful country also lies in the Cumberland Mountains that extend through eastern Kentucky and Tennessee.

In mid-continent, among the prairies and plains, you are not likely to look for mountains. But there are ranges and plateaus that rise to creditable heights and have distinctive structures, splendid views, unusual combinations of wildlife, and mountain dwelling people who retain some of the pioneer spirit of those who crossed the Appalachians and explored the West.

Most extensive of these interior highlands are the Ozark plateaus of southern Missouri and northern Arkansas and the Ouachita Mountains of eastern Oklahoma and central Arkansas. Between these highlands the Arkansas River flows through a broad east-west valley characterized by fertile farm lands.

The Ouachita Mountains are mature folded mountains similar to the folded Appalachians. The range is some 250 miles long, and while the topography is moderate, altitudes go up to 2,850 feet. The original rocks were developed in sedimentary layers to a depth of some 25,000 feet from Cambrian to Pennsylvanian time. At the end of the Pennsylvanian period the strata were highly folded and faulted. Some layers also were overthrust to the north as much as twenty miles. The folds and valleys trend east and west. Beneath the Arkansas River Valley coal measures are found.

The Ozark Plateaus are less folded than the Ouachita Mountains, although they are built of deep water deposits of the same age. A sharp rise at the southern edge of the plateaus is called the Boston Mountains. North of this rise the strata are almost horizontal. The western portion is called the Springfield Plateau; the eastern portion is the Salem Plateau. In the northeastern portion of the Ozarks granite appears at the surface and forms St. Francis Mountain (1,500 ft.).

The Ozark region of Missouri and Arkansas contains many noted springs, especially in areas underlaid by limestone. The flow of Mammoth Springs, Arkansas, has been estimated at 100,000 gallons per minute. Van Buren Spring in Missouri produces even more water. Hot Springs in Arkansas is famous for its curative waters and is the site of our first National Park.

The Ouachita National Forest and the Ozark National Forest embrace thousands of acres of mountains, ridges, lakes, and streams in this land of forests, shrubs, and flowers. In addition to construction timber the harvest from these mountains includes pulpwood, staves, ties, poles, veneer, logs, lead, zinc, and other minerals. The varied flora, abundant wildlife, and clear water make these highlands attractive for hikers, rock hounds, campers, hunters, fishermen, photographers, bird watchers, artists, and outdoor families. Talimena Skyline Drive follows the lofty crests of Rich Mountain and Winding Stair Mountain for fifty-five miles between Mena, Arkansas, and Talihina, Oklahoma. It affords extensive views of the Ouachita Mountains.

In southwestern Oklahoma, west of Fort Sill, the Wichita Mountains emerge from the grasslands where millions of bison used to roam. Bison are still there, about 1,000 of them, protected in the Wichita Mountains Wildlife Refuge. Fifty other species of mammals, including opossums, shrews, armadillos, badgers, elk, and whitetail deer, have been listed for the refuge.

But the last time I was there, the prairie dogs had vanished and attempts to transplant them had ended in failure. The varied habitat, ranging from prairie to forest, and the southern location of these mountains, supports reptiles, fish, and mammals from east and west. Birds from north and south live together or meet at the refuge on spring and autumn migrations.

The Wichita Mountains, like the Ouachita Mountains to the east, are faulted and folded. The uplift, some 300 million years ago, exposed lavas and sandstone overlaid by limestone. Subsequent erosion stripped much of the material from the mountains and deposited it where the modern prairie now lies at the edge of the mountains. The tallest peak now remaining is Mount Scott (2,464 ft.). At lesser elevations, rocky ravines, canyons, and cliffs, supporting a mixture of oaks and other trees, give refuge to deer, eagles, rock wrens, sparrows, hawks, and wild turkeys.

In western South Dakota and northeastern Wyoming the Black Hills are a classical example of dome mountains. They were formed by thick magma rising beneath overlying layers of limestone and sandstone. Granites have been exposed by erosion of the central mountain mass. Harney Peak (7,242 ft.), the highest summit east of the Rocky Mountains, is composed of pink granite and is a picturesque feature of the oval shaped mountain range which is one hundred miles long and fifty miles wide. Surrounding the central Precambrian rocks are a series of exposed strata composed of Paleozoic limestone, Cretaceous sandstone, and Tertiary sandstone. Thus the rocks are progressively younger with increased distance from the center of the Hills.

The Black Hills receive more rain than the surrounding plains. This results in an interesting concentric zonation of vegetation beginning with the grasslands at the lowest elevations and progressing through shrublands to pine and spruce on the slopes and in the rugged canyons. Trees and other plants also find a meeting ground from many directions in the Black Hills. Paper birch, white spruce, and aspen have their affinities in Canada; ponderosa pine and lodgepole pine in the Rockies; bur oak, elm, and ash from the eastern deciduous forests; grasses from the prairies, and sagebrush from the western deserts.

Because of the multiplicity of habitats produced by varied vegetation communities and altitudinal differences, a corresponding variety of birds and animals occur in the Hills. The varied wildlife, gentle mountain topography, scenic views, gold mining history, good fishing, and mild summer climate make the Black Hills attractive for multitudes of recreationists.

In the treeless prairie near Bottineau, North Dakota, lies a hilly section on the border between the United States and Canada known as the Turtle Mountains. These hills reach altitudes between 2,000 and 3,000 feet above sea level—only a few hundred feet above the surrounding plain. They are called mountains because they stand so prominently above the prairie. The Indians called them Turtle Mountains because of a fanciful resemblance to that sacred animal. The hills cover about 1,200 square miles in North Dakota and some 400 square miles in Canada where part of their area has been designated a Provincial National Park.

The Turtle Mountains originally were a part of the Missouri Plateau but erosion separated them from the western highlands. During the Ice Age glaciation rounded the hills and left hundreds of feet of drift material as the ice retreated. Now the wooded hills, the numerous fresh water lakes, and the great variety of wildlife make this island in the prairie a favorite hunting, fishing, and recreation site with a mountainous atmosphere cherished by outdoor people.

Elk, moose, and woodland caribou roamed these hills before the coming of white men. Now there is a fair abundance of whitetailed deer. Beaver, muskrats, grebes, herons, ducks, and gulls add interest to the lakes and marshes. The ruffed grouse strums in the woodlands and prairie chickens boom on their dancing grounds in the grasslands.

At the western edge of the plains we come to the great confrontation of the Rocky Mountains. Like a giant wall the rocks ascend tier on tier until dozens of their peaks top the horizon at more than 14,000 feet. These are new mountains, barren above timberline, deeply sculptured by glacial action in the north, untamed by erosion, fractured by faults and uplifts, and dissected by deep canyons with precipitous cliffs. Time has been too limited for the Rockies to be rounded into gentle hills and rolling summits and clothed with forests to the tops of their ridges as have the Appalachians. But still, the Rockies are impressively beautiful with their evergreen forests, intermontane grassland parks, rushing streams, altitudinal climates, and zones of plant and animal life.

The Rocky Mountain chain is the longest on earth. Beginning in Alaska it extends through Canada and the United States and ends in the Sierra Madre Oriental in eastern Mexico. In the United States the Rocky Mountains comprise three provinces: the Northern, Middle, and Southern Rocky Mountains. While each of these provinces differs in its land forms, the system has many features in common: great relief, with many summits more than a mile above the mountain bases; diversity of rocks, varying from igneous and metamorphic to sedimentary; extensive forests with relatively few species; great mineral wealth; and a source of water that supplies the needs of one fourth of the nation.

The Southern Rocky Mountains lie mostly in Colorado with extensions northward into Wyoming and southward into New Mexico. The eastern ranges include: the Laramie Range, Medicine Bow Range, and Sierra Madre in Wyoming; the Front Range lying next to the Great Plains from Cheyenne, Wyoming to Pueblo, Colorado; and the Sangre de Cristo Range, extending to Santa Fe, New Mexico. On the western slope of Colorado are the Gore Range, Park Range, Sawatch Range, Elk Mountains, the San Juan Mountains in southwestern Colorado, and the Jemez Mountains in New Mexico. In addition to these, fifty or more mountain ranges have been named in Colorado.

In the Southern Rocky Mountains hundreds of challenging peaks have summits more than 10,000 feet above sea level. Colorado, with more high mountains than any state except Alaska, claims 53 peaks that rise to 14,000 feet or higher. Among the notable ones are Pikes Peak (14,110 ft.), west of Colorado Springs; Longs Peak (14,255 ft.), in Rocky Mountain National Park west of Estes Park; Mount Evans (14,264 ft.), in the Front Range and with the world's highest automobile road reaching to the summit; and Mount Elbert (14,433 ft.), Colorado's highest peak. In the Collegiate Range, north of Salida, Colorado, some of the peaks have familiar names: Mount Harvard (14,420 ft.); Mount Columbia (14,073 ft.); Mount Yale (14,196 ft.); Mount Princeton (14,197 ft.); and Mount Oxford (14,153 ft.). Despite their heights, most of the peaks in the Collegiate Range

Alpine vegetation in the foreground and glacial cirques between the peaks that rise 13,000 feet in the background in Rocky Mountain National Park.

are easy to climb. Some can even be managed most of the way with a jeep.

Most of the summits of the highest peaks in the Colorado Rockies are Precambrian rocks from which the sedimentary layers were stripped by erosion after their uplift from the inland sea which extended from the Arctic Ocean to the Gulf of Mexico. The folding and uplifting of the mountains finally expelled the sea and the younger strata were turned up like ski tips against the flanks of the Precambrian core rocks. The upturned edges of these sedimentary layers form the picturesque hogbacks at the edge of the plains north of Denver. In addition to uplift forces, many of the Rocky Mountains were formed by volcanic activity. The Spanish Peaks south of Pueblo are ancient volcanic peaks. Volcanism was extensive in the San Juan Mountains. And volcanic activity still continues in the Yellowstone Park region.

The Middle Rocky Mountains are a diverse group of uplifts lying in Wyoming and Utah and forming a semicircle including the Bighorn, Owl Creek, and Wind River Mountains, the plateau at Yellowstone Park, all in Wyoming, and the Uinta Mountains and the Wasatch Mountains in Utah. The Bighorn, Owl Creek, Wind River, and Uinta Mountains are uplift mountains;

the Wasatch Mountains are fault-block mountains; and the Yellowstone area is a huge lava plateau. Adding to the diversity is the east-west trend of the Uinta Mountains. Many of the mountains in this province are high and impressive, especially the Teton Range and the Absaroka Plateau in northwest Wyoming.

The Tetons, towering above glacial lakes, exhibit fine examples of precipitous peaks and glacier carved U-shaped valleys. The eastern front is spectacular since it is a fracture plane that rises as a sheer wall above the Jackson Hole country. The tilted block of the Tetons slopes to the west and from that direction the approach is gentle and not impressive.

Westward from the southern Big Horn Mountains a low range of block mountains extends to the vicinity of Dubois, Wyoming. The Wind River Canyon, with steep walls towering more than 3,000 feet, divides this range into the Bridger Mountains to the east and the Owl Creek Mountains to the west. A modern highway now traverses the canyon. Formerly the area was impassible to wagons. Freighters had to use other passes to reach the Bighorn Basin.

Among lesser known mountains are the Absaroka Range in northwest Wyoming and the Beartooth Mountains which slope upward into Montana where Granite Peak (12,799 ft.) is that state's highest mountain. Timberline lies at 11,000 to 12,000 feet; above this is some of the finest alpine country in America where flowers grow with incredible beauty, the sky is dazzling blue, and there is no deadening air pollution. The landscape is not one of peaks but of bogs, lakes, streams, and deep glaciated valleys carved from vast horizontal sheets of volcanic debris.

There are many more attractive mountain ranges north of Wyoming in the Northern Rocky Mountains that continue across Montana and Idaho into Canada. Some are rounded hills, such as the Okanogan Highlands west of Spokane, Washington, which are less than 5,000 feet in altitude. Eastward, the mountains are blocks of the earth's crust that have been thrust upward between faults—the geologists call these horsts. The valleys are depressions between parallel faults in the earth's crust—these are called grabens.

In central Idaho one of the world's largest batholiths, a granitic uplift outcropping over thousands of square miles, forms the Clearwater and Salmon River Mountains, and extensive dissected plateau areas where the valleys have been eroded to depths of 3,000 feet. In Glacier National Park the geology is exceptionally interesting because of billion-year-old rocks lying on top of young Cretaceous sedimentary rocks. Apparently the old rocks slid downhill over the young rocks. Then the enormous slab was tilted upward so it now slopes to the west. Glacial action during the Ice Age carved massive valleys and produced beautiful lakes where their moraines impounded the mountain waters.

The Northern Rockies have a wealth of forests, wildlife, and scenic beauty. Ponderosa pine, lodgepole pine, larch, western white pine, and Douglas fir have great commercial value. Alpine meadows come down to 6,000 feet. Deer, elk, bear, mountain sheep, and mountain goats are among the best known large mammals observed. Mountain flowers make the slopes and meadows a botanist's paradise in late spring and summer.

The Cascade Mountains and the Sierra Nevada comprise one of the great highlands of America. The Cascades begin in southern Canada and extend to the Feather River in California. From here to the Mohave Desert the Sierra Nevada forms a mighty barrier, 400 miles in length with summits exceeding 14,000 feet. The east side of this enormous mountain front borders the arid country of the Columbia Plateau and the Great Basin. The panorama of these majestic mountains varies from glaciated valleys and snow capped volcanoes in the north to precipitous escarpments and high granitic pinnacles in the south.

The Cascade Mountains are notable because of their classical volcanic peaks. Among these are Mount Baker (10,778 ft.), Glacier Peak (10,568 ft.), Mount Rainier (14,410 ft.), and Mount Adams (12,307 ft) in Washington; Mount Hood (11,235 ft.), Mount Jefferson (10,499 ft.), and the Three Sisters in Oregon; and Lassen Peak (10,457 ft.) in California. These volcanoes originated in part in post-glacial times. Consequently, they are relatively uneroded and retain their symmetrical forms.

Glaciation has produced many beautiful lakes in the northern Cascades. Many of these, such as Spectacle Lake, Fawn Lake, and Cathedral Lake are deep in the mountain wilderness of northern Washington and are accessible only by foot trails or with horse pack trains. The most spectacular one is Lake Chelan, formed by a gigantic glacier that deepened the Chelan River to below sea level and filled the lower end of the valley with a high moraine. The lake is approximately a mile wide and winds westward among increasingly high mountains for some sixty-five miles. Facetiously, I once told a lady who had taken the tourist boat to the upper end of the lake, that the water there was 1,000 feet higher than at the lower end of the lake. Looking at the impressive mountains rising from the water's edge, she momentarily believed me. All was forgiven when I guided her to the colorful Rainbow Falls, 312 feet high, a short distance above the head of the lake.

A multitude of snowy peaks and glacier capped volcanic cones enhance the alpine scenery in the Cascades. Mount Adams supports nine glaciers and Mount Rainier has twenty-seven. Since Rainier stands apart from the Cascade range and rises from near sea level to 14,410 feet it is one of the dominant mountains in America. The giant trees at its base, the vivid wildflower meadows, the alpine flora, and the perpetual snow and ice near the summit present climbers with a panorama of life zones equal to a 2,000 mile south to north horizontal journey on the lowlands.

Higher than the Cascades, the Sierra Nevada forms a mighty mountain barrier between the Great Basin and the great Central Valley in California. The Sierra is a great fault-block tilted high at its eastern edge and sloping westward under the Great Valley. The eastern edge of the block is 6,000 to 11,000 feet high with peaks rising to more than 14,000 feet. As with other mountain ranges the Sierra has undergone repeated episodes of deposition, uplift, folding, intrusion of granite from deep in the earth, metamorphosis of its rocks, and repeated volcanic activity.

The Sierra offers many things to many people. Mount Whitney (14,494 ft.) is the highest mountain in the United States outside of Alaska. Other giant peaks exceed 14,000 feet and mile-deep canyons add to the superlative scenery. Rushing rivers please hordes of fishermen and supply water to towns, cities, and to farmers for irrigation. Gold mining, beginning with the placer mining in California from 1849 to 1860, followed the discovery of the precious metal in river gravels. The gold had been carried by water from rock veins developed during the Pennsylvania and Jurassic periods. The universal and eternal processes of erosion revealed them to human eyes.

Along the Pacific Ocean, from northern Washington to southern California, the Coast Ranges lie west of the Puget Trough and the California Trough which separate them from the Cascade-Sierra Mountains. The Olympic Mountains and the Oregon Coast Ranges are separated from the California Coast Ranges by a geologically unrelated range, the Klamath Mountains. In age the Klamath Mountains correspond more closely with the Sierra Nevada and with the Blue Mountains of eastern Oregon. In the Triassic period, nearly a quarter billion years ago, sediments were laid beneath the sea in the Klamath region which now consists of uplifted igneous masses.

The Olympic Mountains occupy the northwest corner of Washington, north of Olympia and west of Seattle. These craggy mountains, topped by Mt. Olympus (7,954 ft.), date from a Pliocene uplift and because of their youth, ten to fifteen million years, are only slightly eroded. From the central cluster of peaks and ridges, valleys and dashing rivers radiate in all directions. These rivers, fed by glaciers, snow, and high rainfall, flow from the high country to the Pacific Ocean on the west, the Strait of Juan de Fuca on the north, and the Hood Canal on the east.

The names of these rivers sound unreal to people who are not Northwesterners: Humptulips, Queets, Hoh, Bogachiel, Soleduck, Pysht, Elwha, Dungeness, Dosewallips, and Hamma Hamma. These and other rivers carry America's greatest precipitation—150 inches or more each year—through rainforests of giant Douglas firs, Sitka spruces, western red cedars, and western hemlocks that nearly match the sequoias and redwoods of California in height and diameter. No subalpine country in America is more scenic than the flower covered meadows in the high Olympics which are whitened by more than fifty glaciers. The Hoh glacier on Mount Olympus is more than three miles long with ice reaching a depth of 900 feet.

The Oregon Coast Range extends from southwestern Washington to the Rogue River in western Oregon. Most of the summits are less than 3,000 feet and the tree covered slopes resemble those of the Adirondacks and some of the Appalachian plateaus.

Neahkahnie Mountain, which you can traverse on the Oregon Coast Highway, rises 1,700 feet directly above the sea. Nine hundred feet of this is an almost vertical cliff that stands as a fault along the coast line. Although the sea has worked its way on the sandstones of the Coast Range, and has left rock spires or sea stacks in the water by its landward progress, it has not crumbled the hard igneous mass of Neahkahnie Mountain.

The Coast Ranges in California begin near Eureka, south of the Klamath Mountains. This nearly 400 mile long series of ranges that lie parallel to the Pacific Ocean ends near Point Conception. Few of these mountains exceed 2,000 feet in elevation in the northern regions. North of San Francisco four ranges, the Marin, Sonoma, Napa, and Yolo, are separated by valley faults. The Berkeley Hills east of San Francisco Bay are a part of the Coast Ranges.

South of Monterey Bay the Santa Lucia Range emerges abruptly from the ocean front where the Big Sur provides some of the most pleasing scenery on the continent. Of special interest to people living in the Coast Ranges is the San Andreas fault which had a displacement of approximately twenty feet during the 1906 earthquake. The great earth block to the west has been

Joshua trees and other yuccas grow at the base of desert mountains in eastern California. Vegetation on the rocky slopes is sparse owing to shallow soil and the arid climate.

moving northward since Cretaceous time. If the tectonic experts are right, the Los Angeles vicinity will ultimately be opposite San Francisco, not in our time but possibly 50,000 years in the future.

From Point Conception a series of mountains known as the Transverse Ranges trend east and west. These are much faulted and folded mountains, some with elevations exceeding 10,000 feet. Well known among these are the San Gabriel and San Bernardino Mountains. Notable peaks are San Gorgonio (11,502 ft.) and San Antonio (10,064 ft.), locally called Old Baldy. West of Los Angeles are the Santa Monica, Topatopa, and Santa Ynez Mountains. The Channel Islands are the exposed peaks of the western extension of the Santa Monica Mountains. All of this region is characterized by active faults, especially the San Andreas and the San Jacinto faults.

In extreme southwestern California, mountains of the Peninsular Province lie south of Santa Ana and the Coachella Valley, west of Imperial Valley, and southward for 800 miles where they form the backbone of Baja California. These include the San Jacinto, Santa Rosa, Santa Ana, Vallecito, and Laguna Mountains north of the Mexican border. Along the coast, terraces cut by ocean waves during Pleistocene time give evidence of uplift of at least 1,300 feet above sea level. When we consider the active faults, the earthquakes, the volcanic activity, and the rising shoreline, we can believe that the Pacific Mountain System is still growing and that the restless crust of the earth is still moving in this part of the world.

Grand Mesa in western Colorado rises to 10,000 feet. From Land's End in the distance one can look down to the Colorado River, 5,000 feet below, and the Book Cliffs on the right and to the Gunnison River and the Uncompahgre Plateau to the left. Grand Junction is named from the place where the two rivers come together.

3

Mountain Landscapes

MOUNTAINS HAVE so much variety that no person can experience all their moods, their diversity of form and climate, or their multitudes of environments for living things. One can pit his strength against their rugged shoulders for forty years and not exhaust appreciation of their majesty, mysteriousness, complexity, and intrinsic beauty. Each sojourn into the high hills can give new experiences, unexpected adventure, and glimpses of beauty not seen in the more gentle terrains where most of us live.

The enjoyment of mountain scenery—the broad view of landscapes—is sufficient compensation for many people. Topographic variety in the mountain world is unlimited. Look at a mountain from any vantage point, then step aside to another viewpoint, and the whole scenic composition changes. In this respect, scenery is like art, the more we study it the more we see in it to enjoy.

The types of landscapes are so varied that anyone can choose his favorite, enjoy its configuration, and maybe understand its origin, what it is now, and what it will be. Where you are is not too important. On the grand scale, for example, there are vast blue distances in Shenandoah National Park where the folds of the Appalachians stretch to the far horizon. There are rounded crests in the New England-Adirondack mountain regions where huge ice sheets thousands of feet thick once abraded the valleys and summits. Now these are clothed with colorful forests interestingly varied by striated bedrock outcrops on ridges and in notches where highways traverse the scene.

For New Yorkers there are the Hudson Highlands where the river flows between the Catskills on the west and the ancient Taconic mountains on the east. The lava columns of the Palisades rise abruptly from the river to noble heights, having been freed by erosion from encompassing sandstones. If one speculates long enough about these mountains, or any other mountain scenery, the fact of erosion becomes apparent as one of the prime forces in the development of mountain grandeur. Even the Old Man of the Mountains in Franconia Notch, New Hampshire, looks down at Profile Lake, because erosion chiseled him out of Cannon Mountain in the ancient past.

In the West the stark ramparts of the Rockies loom in castellated ranks above the plains inviting the visitor to come within their ranges to see peaks, cliffs, valleys, canyons, and parks. The serrated battlements of the Sierra Nevada with their cirques, domes, spires, and rock fans are the very embodiment of the raw materials of earth, sculptured by water, wind, ice, and gravity, and exhibiting scenic splendor that the artist can only copy but not improve.

In the volcanic country from northern California to northern Washington, the great snow covered, symmetrical volcanic cones rise above the surrounding mountains with inspiring majesty. From vast distances you are impelled to look at them repeatedly because they are the most visible and definitive features in the landscape.

But even the most casual observer can be aware of some of the natural processes and the tools of nature that have created and are still changing the land forms of the continent. Running water is one of these. The weather brings water to the slopes, meadows, and valleys in the mountains where it falls as rain or accumulates as snow in winter. There water seeps into the joints and cracks in rocks, dissolves minerals, or expands as ice and chips off particles of stone. There mountain springs and streams also are born, and through their action carry the earth materials from high to low places and to the sea from which new mountains may eventually be born.

These tools of nature—water, wind, ice, gravity, the chemicals produced by roots of plants—work at different rates on different

Vertical cliffs are formed by layered rocks with joints and fractures. Alternate freezing and thawing loosens the rocks which fall and produce a talus slope at the base of the cliff.

kinds of rocks. Mountain peaks usually are made of hard rocks and thus persist, while the softer rocks that surround them are eroded into lesser mountains and foothills. Mesas, those flat-topped hills with vertical sides, exist because hard rocks cap the more easily eroded rocks below. Scenic monuments and spires in western canyon lands sometimes represent the remnants of mesas. Their grotesque shapes are produced by differential weathering of alternating layers of hard and soft rocks. Their sharp profiles, so characteristic of dry climates, stand in marked contrast to the rounded peaks and hills of the Adirondacks and Appalachians where great age of the mountains, moist climates, and thick soil have produced gentle profiles.

Landscapes always are more meaningful if one knows something of their geological history and the forces that produce basic land forms. As a beginning, remember that the rock layers of all mountains were originally deposited beneath the sea. The mountains themselves are uplifts of the earth's crust. If the uplift is even and the bedrock remains flat, the landscape is likely to be a plateau, with canyon lands, escarpments or rims and outlying mesas for its topography. If the bedrock is tilted in one direction the uplift results in a giant fault-block such as the Sierra Nevada, steep on the east side and gently sloping on the west side. If the uplift is faulted and folded the resulting mountain range is a complex of peaks, ridges, and valleys in many patterns. The local landforms within these landscapes are creations of erosion, deposition, glacial action, solution of rocks by underground waters, or thermal actions where the final vestiges of volcanism remain.

Among the landscape features, not consciously recognized by all visitors in the mountains, are the great valleys that have been remodeled by glaciers. In the Olympic Mountains in Washington, in the northern Rockies, and in Alaska, glaciers are still wearing away valley walls, excavating valley floors, and depositing moraines of rock debris. But much of the glacial topographic variety we now see is caused by ice that flowed from the heights in ages past. When the climate warmed and the ice melted, the cirques, rock troughs, hanging valleys, and rock moraines were exposed to view.

Cirques, the blunt or rounded valley heads in high mountains, are widely distributed. The Great Gulf Wilderness near the top of Mount Washington is a deep glacial valley now filled with trees. Tuckerman's Ravine, on the east slope of Mount Washington, also is a glacial cirque. From the alpine slopes in Rocky Mountain National Park you can look down into Forest Canyon where glacial lakes, or tarns, now occupy the valley floor that once was scraped clean of rocks. Elsewhere the evidence of ferocious ice carving of mountains is almost universally present in the Rockies, Sierras, and Cascade Mountains.

In the beginning many of the western mountain valleys were V-shaped as a result of stream erosion. Then when the glaciers appeared, the rock studded ice chiseled and gouged the valley sides and ground of the side spurs until the canyons became U-shaped. Yosemite, Kings, and Kern canyons in California were formed in this way. In some instances the main canyons

The majestic Northern Rocky Mountains in Glacier National Park present a mixture of alpine scenery, glaciers, forests, and lakes. The pyramidal shape is the result of past glacial action. In the famous Lewis overthrust the older mountains were shoved some thirty miles eastward over younger rocks, thus reversing the usual order of strata in the earth's crust.

were gouged more rapidly and deeply than the tributary canyons. When the ice melted the side valleys were left hanging above the main valleys. From these perched valleys streams now cascade in spectacular waterfalls. At the maximum of glaciation the ice, thousands of feet deep, even helped to carve Matterhorn-like peaks. The Grand Teton in Wyoming is an example of how the inexorable creeping tongues of ice carved the pyramid shape of this beautiful mountain.

The moraines or piles of rock debris dropped by melting glaciers contribute much to scenic variety. In some instances the ice that accumulated around peaks expanded to form a piedmont or sheetlike glacier on the lower flanks of the mountains. The rock debris picked up by the ice filed holes in the floor beneath the glacier. In the Wind River Mountains of Wyoming, for example, these holes are now marked by dozens of small lakes lying between 8,000 and 9,000 feet elevation. Lakes also are formed where the rate of ice advance and melting were equal and the rock debris was deposited in one place to form a dam or terminal moraine across the valley.

Canyons are now among the most impressive features of western mountain and plateau lands. They are notches in the earth, not cracks or faults, made by the rivers they contain. The Grand Canyon, as most people know, is so deep that the Colorado River appears as a thin ribbon of water in the depths of the earth. The canyon itself has been widened through aeons of time by erosion from side tributaries, and by crumbling of the cliff walls as rock materials have been loosened by frost, water, wind, and gravity. The stair-step profile has been formed by alternating layers of hard and soft stone. The vertical cliffs are hard rocks from which material has fallen to make talus slopes that extend to the edges of the next cliffs below. At the bottom of the canyon the river still is cutting a narrow groove through some of the most ancient rocks on earth.

The grandeur of the canyon stems from its location in an arid land where erosion proceeds at a rapid rate in contrast to the Appalachians where forest covered slopes and deep soil retard disintegration of the rocks. Great age also has permitted rounding of the folded and faulted eastern mountain tops and ridges. In the western plateau country the nearly level rock strata have been slowly uplifted while the muddy river scoured its bed at the same rate, thus maintaining its low level altitude. For those who are willing to walk or to ride mule back into the depths of the Grand Canyon there is no place on earth where geological history is more clearly exposed.

A similar descent through time and climate can be a great adventure in Hells Canyon, that mighty gash made by the Snake River between Oregon and Idaho. The Seven Devils range rises on the east side of the gorge and from the top of He Devil Peak down to the river is 7,900 feet. This is 2,250 feet deeper than the Grand Canyon. On the west side of Hells Canyon, Hat Point (6,984 feet) can be reached by airplane on the small landing strip near the Forest Service guard station. Or one can drive on a steep rough road from Joseph, Oregon, along 2,000 foot precipitous grassy slopes.

This is enchanting country. In the coniferous forests near the rim of the canyon are deer, elk, bear, cougars, coyotes, pine squirrels, Stellar's jays, and Clark's nutcrackers. The blue grouse are so tame they hardly move from underfoot as you wander through the forest. On a pack trip down Saddle Creek to the bottom of the Canyon, north along the Snake River, and back up Temperance Creek, I saw seven porcupines, more than fifty deer, and two black bear. My most hair-raising experience came as the pack train was traversing a narrow ledge along the canyon wall with no place to go but straight ahead. As we looked down our right saddle legs at the river far below, a rattlesnake slithered across the trail beneath my horse's feet and coiled in a clump of grass beside the trail. The snake did not rattle and the mule pack train behind me passed without incident. Gradually my hat settled down on my head.

In the Snake River, which is only 100 feet wide in narrow places, white sturgeon weigh up to 300 pounds. Fishermen on the Idaho side used to pull them out with a horse attached to a strong fishing line. At low elevations near the river, blue grama, sand dropseed, and other grasses characteristic of the Great Plains hundreds of miles to the east remind one of the semi-arid climate of the short-grass prairie. Even prickly pear cacti grow vigorously in the rocky soil.

Few trees grow in the lower slopes, excepting those that flourish beside the river itself. Far above, vertical walls of basalt ascend in stair-step fashion, alternating with steep grass covered benches that slope precipitously toward the next lower cliff. Cattle graze here by following the contoured cat-steps or paths made by many generations of livestock. When ice is on the canyon sides the animals sometimes fall over the cliffs. There is a story of a man who visited at a family cabin in the canyon and noticed a human skull on a shelf. "Who was that?" the visitor asked. "Oh, that was our Pa," one of the boys replied. "He and his horse fell off the rocks in a snow storm and we didn't find him until a long time afterward."

An even more frightening chasm is the Black Canyon of the Gunnison River in Colorado. The Painted Wall is a sheer 2,250 feet from rim to river. When one looks down to the river, past the pinnacles that rise from the canyon's floor, or creeps to the edge of either the north or south rim, the view into the dark depths gives one the cold shivers. In spite of its depth the width in places is hardly more than one quarter of a mile. In the wild desolation of its depths mighty rock piles and the cascading river make the trip clear through the canyon an almost impossible task. Few men have succeeded, even with rubber boats and rock climbing equipment. At the mouth of the forbidding canyon, west of Cimarron, Colorado, I used to find fabulous trout fishing in the gloomy canyon.

Quite the opposite of canyons are numerous gigantic rock skeletons that tower above the landscape where erosion has removed the surrounding material from the cores of volcanoes. The lava neck of the volcano then remains as a towering monument, sometimes with the internal plumbing of the volcano remaining in the form of radiating dikes. Ship Rock in New Mexico is a notable example of a volcanic skeleton some 1400 feet high. Devils Tower, visible for dozens of miles across the flat Wyoming plains, also is a pillar of lava nearly 1,300 feet high. Close by, it is seen to consist of vertical columns of igneous rocks separated into long prisms with conspicuous joints. Prisms formed from lava flows tend to be hexagonal and may be 500 feet long. The Devils Postpile National Monument in California has a remarkable display of this type of basaltic column formed by ancient lava flows.

Volcanic landscapes also occur in the form of jagged lava fields and cinder cones such as may be seen in the Craters of the Moon in Idaho and on the eastern slopes of the Cascades in central Oregon. Lava Butte, a few miles south of Bend, Oregon, for example, is a perfect cone a few hundred feet in height. It is easily climbed by automobile on a spiral road that stops at the edge of its crater in the summit. A few miles to the east are lava gardens dotted with mighty "post holes," formed as the hot lava cooled around ancient ponderosa pine trees. In this same vicinity are ice caves and lava tubes, yards wide and high and of unknown length, where molten rock once flowed beneath the surface in the direction of the Deschutes River. Unlike limestone caves, these lava tubes have not produced stalagmites or stalactites.

Natural bridges are among the most unusual of the erosion phenomena in the sandstones of the western plateau country. A natural bridge results when a deeply entrenched and meandering river or stream cuts through the divide between loops and forms a hole through which the river flows. At first the hole is small and the bridge abutments are massive. As erosion and weathering enlarge the hole the bridge becomes more slender and delicate.

Balanced rock near Colorado Springs, Colorado. Differential erosion removes the softer rocks faster than it weathers the harder portions. Formations like this occur when the hard and soft layers are close together.

Finally, when the center span falls, only the rock buttresses remain as rock columns or stony knobs. One of the best known natural bridges is Rainbow Bridge. Other beautiful examples in Natural Bridges National Monument in Utah are Sipapu, White River, and Kachina Bridge. The latter has a span of more than 200 feet and is 108 feet from the stream bed to the base of the arch. Natural bridges, like all other things in nature, eventually erode, fall, and disappear.

Arches and rock windows have an origin different from that of natural bridges. Arches are formed by erosion of narrow rock walls which have been uplifted and cracked apart along vertical joints in sandstone. Water and frost chip away the surface of the wall in thin spots until a window is exposed. As weathering continues, the hole enlarges and eventually the arch appears. In much jointed rock the process is more rapid if the underlying layers of sandstone are softer than the cap rock. As with natural bridges, buttresses, spires, or stone monuments are left when the arch collapses. There seems to be no end to the variations in size, shape, and ornamentation of these erosional forms that result from the integrated movements of the earth's crust and the processes of weather that never cease.

The mountain world has many micro-wonders that are as small as the stones beneath our feet. A pebble, for example, that has been long in place may be faceted by wind driven sand. On mountain tops, rock particles driven by the prevailing winds grind the surfaces of pebbles and rocks smooth. If the pebble gets turned it presents another surface to the wind and another facet is formed. Some of the faceted gravels deposited at the end of the Ice Age in New England indicate that strong winds were prevalent in those ancient times since the grinding process is not now in progress.

Most of us know how wave worn beach cobbles become rounded like balls. The constant tumbling in the surf wears away the rough edges and corners of stones. These cobbles once were blocks of stone cracked from rock layers by stresses and strains. The cobbles in rushing mountain streams are similarly rounded and polished by tumbling when storms send flood waters rushing downstream. Mountain rivers also have other sizes of rocks— boulders, granules, sand, silt, and clay. These ultimate particles, silt and clay, when mixed with organic matter from plants and animals, form the soil of meadows, flood plains, and even the land beyond the mountains. Some are carried to the sea to be deposited in layers where they may become rocks again and in some future aeon may be lifted once more to form mountains.

There are ways to see mountain landscapes in action. Fortunately, or unfortunately, depending on one's point of view, most of us will never experience the greatest violence that nature can produce, a volcano in catastrophic eruption, a shattering earthquake, or a flood that scours away half a state when an inland sea such as ancient Lake Bonneville cuts through its margin and changes the course of a Columbia River. Happenings like these, when the earth was an unquiet land, have given us some of the wonders we now see in peace and quiet. But in a smaller scale one can still see or experience the beauty, the dynamics, and even the savagery of the earth in action.

One way is to live in a mountain cabin and watch the seasons

The great Tillamook burn in the Coast Range of western Oregon missed this virgin stand of giant Douglas fir trees. Loggers have since cut this remaining stand of trees.

go by where high altitude gives them special character. The snow storms are magnificent. Snow avalanches clear great paths on forested slopes. Then as spring comes up the mountain the kinnikinnick shows its green shrubbery, the paintbrushes, yarrows, and dandelions burst forth, ferns thrust up their fiddle heads, and the migrating birds appear. In full season the firs and spruces sprout new leafy candles and the resident animals—chipmunks, martins, coyotes, deer, elk, bear—are active in the highlands. When autumn leaves suffuse the landscape with brilliant colors, the summer resident birds depart and new migratory birds appear. Then as the frosts and early snows appear the growing season ends and the large grazing mammals begin their drift to lower meadows and forests, thus contributing to the biological change that ever responds to topographical relief.

Another way is to climb mountains, not deliberately to live dangerously, but to satisfy the urge to see high and beautiful places while pitting one's strength against stony nature. There is no better way to learn how solid, and yet how fragile, the rocks can be than to trust one's life to pitons hammered into joints and cracks on the face of a thousand foot cliff. Expert mountaineering requires not only a knowledge of carabiners, ropes, gravity, body manipulation and balance, but of rock structure and the routes for climbing that have been fashioned by geological and physical processes while the mountains were being formed. I confess that I am not an expert mountaineer, but I have had experience with a mountain in action.

It happened on Kings Mountain in the Coast Range in western Oregon. Here the precipitous slopes were denuded of giant western hemlock, western red cedar, and Douglas fir trees in the Tillamook Burn area, when the great fires in 1933, 1939, and 1945 killed an estimated twelve to thirteen billion board feet of timber. Subsequent successional growth of vine maples, salal, salmon berries, black berries, and numerous herbaceous plants permitted reestablishment of the blacktail deer herds. The habitat became so productive for wildlife that rodents, birds, bear, and deer became abundant. Some of the blacktail bucks grew to trophy size.

Making my way through fire-killed snags, decaying logs, head-high ferns, and tangled shrubs I climbed the rugged terrain to the crest of the mountains north of the Wilson River. On a north slope, where the big bucks hide in the daytime, I chose a rock ledge as a vantage point for viewing the mountainside across a steep canyon. For half an hour I never moved. Then with a gentle nudge of my foot I dislodged a stone that went rattling down the canyon, dislodging gravel, stones, and boulders to become a thundering avalanche.

Almost instantly a magnificent buck arose from his bed of ferns across the canyon. His antlers were perfect—no odd numbered or uneven tines as is usual for blacktail deer. He had a total of six points and I estimated the antler spread at close to thirty inches. The blackish forehead, nose, and chest were beauti-

Mount Hood is the favorite Oregon Cascade volcanic cone with perpetual snow near its summit. The author here is hunting deer near Windy Point on the east side of Mount Hood. Timberline occurs here at 5,000 to 6,000 feet.

fully formed. The rich brownish gray of his body and the white belly indicated that he had already developed his winter coat. He looked like more than a 280 pounder, and might qualify for the record books. At the crash of my 30-06 he stiffened for an instant and then gently slumped to the ground. Then his relaxed body began to roll down the canyon side, stirring up gravel, rocks, and boulders, and then vanishing in the avalanche that again thundered down the mountain.

While I stood, listening to the rumble of debris in the rock flume and the attenuated thunder of the avalanche far below, the canyon rim crumbled beneath my feet and I joined the maelstrom in a nightmarish whirl of jabbing, pounding, horrific stream of logging debris, rocks, and mud. For a fifth of a mile, in half consciousness, I grabbed and clutched at embedded stones and exposed tree roots and finally stopped just before the avalanche plunged more steeply into the valley half a mile below.

When three hunters far down the mountainside heard my call and spotted me, they climbed with a rope for half an hour to reach me. A preliminary inventory indicated three fingers broken on each hand from clutching at rocks, a broken jaw bone, compound fracture of my left arm, shoulder and leg injury, and what later turned out to be a detached retina in my left eye. They were so excited at the spectacle of blood and mud they placed the rope around my neck. I suggested that if they could not double the rope under one thigh and over the opposite shoulder to at least put it under my arms. They were good men and they finally got me down. I recovered and went hunting in the mountains the next year. I remember making one comment before we started for the hospital: "Now I have a better understanding of erosion."

The Tillamook Burn in the Coast Range of western Oregon is a prime example of accelerated erosion caused by fire and destruction of the soil on steep mountain slopes. This great burn area which encompasses 355,397 acres is the scene of one of the most famous fires in western North America. The successive fires killed the conifers, removed the soil litter, and increased the sediment load in the Wilson, Trask, Kilchis, Miami, and Tualatin Rivers. The reduced water holding capacity of the charred soils also resulted in faster runoff from rains and cessation of summer stream flow in the tributaries to the main rivers.

Nature began the rehabilitation of the area through the work of bark beetles, fungi, and other agents of deterioration that gradually return dead trees and fallen timber to the soil. At the same time, natural succession produced an abundance of shrubby and herbaceous vegetation while foresters started tree plantings which would require several decades to return the burn to a permanent timber producing area. In the meantime, the mountains are contributing greater amounts of their substance to the sea than they did when the original forest was there. Wherever the surface of the planet is not protected by vegetation or is not solidly bound into hard rock the elements seek out the frailty and begin the destruction and erosion that shapes the face of the earth.

To the casual eye, mountains appear to be stable and indestructible. But we know that they move up and down, that they bend and fold, and that they move sideways along faults. Their steep vertical cliffs and some of their linear valleys suggest that blocks of rock have dropped downward or have been lifted along fault lines. Most of these earth movements are so infinitely slow that we do not see them even in a lifetime of watching.

Steam vents on Mount Lassen in California are active even in winter. This volcanic mountain still has its hot spots and may become active again at some unpredictable time in the future.

However, from study of the great arches in the earth's strata, the classic block-faulted mountains of the Great Basin, the erosion by rivers and glaciers, and the earth scars made by volcanoes and earthquakes, we can infer that we live on a dynamic planet.

Earth's convulsions are visibly in progress in the Montana Rockies and in Yellowstone National Park. Lewis and Clark, on July 4, 1805, heard the sounds of earthquakes when they were at the Great Falls of the Missouri. They likened the sounds to those of a six-pounder piece of ordnance three miles away. The Indians also spoke of the noise that sounded like thunder coming from the mountains.

An earthquake occurred in Montana on December 10, 1872 near Deer Lodge. The ground shook, houses trembled and swayed, and people rushed from buildings. On October 18, 1935, after a series of previous tremors, a fault slippage shook the earth from Canada to Montana, Wyoming, and Idaho, with damage to buildings and other property. Then in the night of August 17, 1959, the Yellowstone earthquake along the Madison River caused a huge slide to dam the river. Some eighty million tons of rock hurtled down the mountainside and up the opposite valley slope.

Many people sleeping in camp grounds were killed and roads in the National Park were blocked by boulders. Geysers in the Park changed their flow, dormant ones became active, and some active ones became dormant. Fissures appeared in the earth's surface near Firehole Lake and quiet water pools turned into geysers. Thus the turbulent Madison Range of the Rocky Mountains still appears to be growing as it has been for fifty million years.

The geological history of Yellowstone itself is long and complicated. More than two billion years ago the Precambrian basement rocks were formed. These were faulted and uplifted to become a plateau when the Northern Rocky Mountains were formed. This plateau became an erosion surface over which volcanic rocks from the Absaroka Range were erupted some fifty or sixty million years ago. Stream and glacial erosion then carved the surface again before the Yellowstone volcanic rocks were erupted.

A long period of volcanism began some two million years ago with catastrophic outbursts of debris. This cycle of volcanism was followed by another about 1,200,000 years ago and by a third some 600,000 years ago. This last outburst may have been one of the greatest eruptions in the earth's history. Hot magma erupting from the depths lifted a great dome through which ash, lava, and gas were blown out. The top of the dome then collapsed into the space left by the escaping magma, thus creating a volcanic crater or caldera many miles in diameter.

There are several calderas on the Yellowstone Plateau which have been partially filled with eruptions from more recent volcanoes. Glaciation in three different periods, between one-third of a million years ago and ending twelve to fifteen thousand years ago, added the final touch to the rock display we now see in Yellowstone. Beneath the surface, molten magma still exists and its heat is the source of the activities of the modern geysers, hot springs, paint pots, fumaroles, as well as the mineral deposits and bright colored rocks that form terraces, pool rims, and geyser basins.

Geysers provide some of the most dramatic displays of earth activity in Yellowstone. Some two million people go there each year to see them. Old Faithful, which has erupted regularly for generations of viewers, is one of the major attractions. Many observers assemble near the geyser tube before each eruption occurs and everyone is impressed by the mighty discharge of water and steam. A few, not counting the geothermal energy advocates, have utilitarian ideas about its possible uses.

One visitor told me he would like to have Old Faithful at his home so he could pipe the hot water under his driveway and sidewalks to melt the snow in winter. A Missouri farmer suggested another use for the geyser: "Think how many hogs you could scald with all that hot water at butchering time," he said. Most people, however, are content to enjoy the sight of geysers erupting in giant columns of water, steam blow-offs, or fountain-like sprays.

Geyser action has been likened to that of a coffee percolator. Surface and underground water accumulates in hot rocks a few hundred feet below the surface. The heat from the magma below is so great that the water becomes superheated in a network of openings similar to the root system of a tree. Being under great pressure the water does not boil but it does expand. The expansion forces water up the tube until it flows out the vent at the surface. Reduction in the weight of the water column relieves the pressure below and the superheated water turns violently into steam. Then the great eruption occurs and continues until the underground channels are clear. Some eruptions last for only a few minutes. Others that occur only at long intervals may last for hours. The rate at which the underground channels refill with water determines to some extent when the next eruption will occur.

Among the less dramatic but still fascinating thermal features of Yellowstone and other volcanic areas are hot springs and fumaroles. These represent some of the dying phases of volcanic activity. Fumaroles are steam vents through which gasses escape through fissures to the earth's surface. The steam frequently contains carbon dioxide, methane, and even iron, copper, or lead ore. If hydrogen sulfide is present a pronounced odor of rotten eggs is present. Bacteria and algae thrive in some of the waters of thermal springs which simply discharge water instead of steam. The waters of some springs also contain mineral matter, such as quartz, which is deposited in a crust called siliceous sinter.

Some of the pools fed by springs are so colorful they exhibit all the hues of the rainbow. This fantasy of color derives from the chemicals dissolved in the water, from substances deposited at the margins of the pools, from algae and bacteria that coat the rocks, and from reflections of the blue sky above. Even the vapors and steam clouds that rise from the hot water are richly tinted because they reflect the blues, greens, reds, and yellows of the pools below.

Mud pots and paint pots are both colorful and vocal. The colors rival the pastels on a painter's palette and are produced by compounds of sulfur, iron, and other minerals. The muds represent the final decomposition of lavas and other materials by acids in hot water and the bubbling, churning, and mixing of particles until they are reduced to clay or mud porridge. The sounds of this activity vary with the consistency of the mud. Some are plops, gurgles, splashes, pops, and hisses while out of the subterranean mud come grumbles, groans, booms and clicks.

There are other strange features in this land of rocks and heat. Each had its birth on one of the stairsteps of time. The travertine terraces made of limestone by Mammoth Hot Spring even have different ages; some are now covered with vegetation, because the mouths of springs have moved, while others are still being deposited. Obsidian Cliff north of Norris Junction is a spectacular outcrop of volcanic glass similar to the obsidian flow in the caldera of Newberry Crater in central Oregon. The basalt columns in Yellowstone River canyon show patterns of cleavage and jointing similar to those found in the Devils Post Pile in California, on the sides of Devil's Tower in Wyoming, or in the Palisades of the Hudson River in New York.

Nowhere can you see as many geological wonders in one locality as you can in Yellowstone country when you stroll among the remnants of earth's history and see geological history still in the making.

Water in the rivers in the Cascade and Coast Ranges in Oregon rushes downward through spruce-fir forests and passes over mountains most of the way to the sea. Salmon climb through the rapids to spawn in gravel beds high in the mountains.

4

Mountain Waters

WITHOUT WATER THERE NEVER WOULD HAVE BEEN MOUNTAINS. When the ancient rocks were deposited as sediments in the sea it was water that brought them there. When the rock beds were uplifted, faulted, and folded, it was water that carved their ancient valleys, chiseled away fragments through freezing or ice action, and returned their substance to the sea. Through the ages it was water that helped level high mountain chains to peneplains or flat topography which later were lifted to new heights so rivers could once more cut channels through their rocks to create new land forms.

Water is everywhere in the mountains. We see it in lakes, ponds, streams and rivers. In the northern mountains it is stored in glaciers and in underground channels from which it gradually creeps toward the sea. It is present in the atmosphere as vapor, and it falls as rain or snow. As liquid in the soil it is absorbed by roots and becomes a part of plants which in turn produce acids and other substances that aid in disintegration of rocks and minerals. Even the rocks themselves contain water bound chemically into their substance. And when a great volcano blows its top and spews lava and ashes over the landscape much of its explosive power comes from water that bursts into steam as pressure above the hot magma in the earth is relieved.

Water is the medium in which great beds of limestone, produced by animals that use calcium in their life processes, is deposited in the sea. Water also is the medium in which sand and gravel are cemented into sandstones that later are uplifted into great plateaus or mountain masses. And water is the principal agent that dissolves limestone and produces caves and caverns with their bizarre deposits of redeposited limestone.

Through earth's history, through all the ages of calm and unrest, water has been a moving force. We now live in a time of relative geological inactivity. It is true that crustal movements and earthquakes still occur. Occasional volcanic activity goes on, and some coastal areas are subsiding while others are rising. But the greatest landscape activity—even though it is hardly perceptible—is water erosion, especially in mountains that have reached heights of several miles.

The birth of some of the most common forms of mountains occurs in the sea. Great depressions in the earth's crust gradually fill with sediments eroded from the land and carried by rivers to the ocean. The weight of accumulating strata of sandstone and other kinds of rocks causes the sea floor to sink. This makes room for more sediments; the final accumulation may be thousands of feet thick. Then forces within the earth cause the sea floor to rise and bring the sedimentary layers to the surface.

Further uplift, flooding, and faulting—possibly caused by tectonic plates crushing together—pushes up mountains. These are made more complex by intrusion of molten rocks into cracks and breaks in the earth's crust. Then rivers begin in the highlands and the universal process of erosion begins to level the land.

People who have traveled through the Appalachian Mountains or have climbed to the top of one of the many parallel ridges in Pennsylvania, Virginia, or Tennessee, must have observed the even skyline along the crests of the mountains. Why are so many of these ridge tops on a common level when their rocks lie at different angles? The answer is that water in the geological past eroded the mountains to a nearly flat land surface called a peneplain. Then the peneplain was uplifted as much as 3,500 feet and a new cycle of erosion began. Under the renewed erosion the hard rocks were left at their present heights while the softer limestones were eroded to a new and lower peneplain.

These skyline peneplains are present in other mountains but are not as easily discernible as they are in the ancient Appalachians. In the West, faulting and folding created the Rocky Mountains approximately fifty million years ago. Then erosion lasting some twenty-five million years practically leveled them, filling their valleys with debris and washing their sediments far out on the Great Plains to the east. Then uplift occurred again and accelerated erosion carved away the softer materials and left the peaks we know today. The many peaks which range from 13,000 to more than 14,000 feet in elevation represent the granite cores the tops of which were once level with the ancient peneplain.

The universal exchange of water between earth and atmosphere is the hydrologic or water cycle. We see evidence of water in this cycle when clouds pass over, when rain or snow falls, when river water flows down hill, when fog develops over the ocean, and when plants and the heat of the sun dry the top layers of the soil. Energy from the sun evaporates water, which becomes vapor in the air and then becomes rain through condensation, or snow by freezing. Precipitation on land then is absorbed into the ground or is powered by gravity to form rills, creeks, and rivers that carry the water back to the ocean. This process of water circulation over the earth now is common knowledge even for high school students. But not all of us associate the shapes of mountains, valleys, and even the great river basins of the continent with the work of water in the ancient past.

Most of the great rivers on the continent have contributed to the making of our landscapes through millions of years. Anyone who sees the Yellowstone river can understand how the swift churning water has cut a deep V-shaped canyon through bright colored lava in Yellowstone National Park. This is a youthful river as evidenced by its magnificent waterfalls, its narrow channel, and its steep gradient. The falls occur where the water flows over layers of hard rocks and drops to layers of softer rocks which erode more easily.

Waterfalls also develop in mountainous areas where the rock strata have been faulted and folded. The streams in these localities, such as in the Sierras, also are youthful streams since they have not had time to wear their beds down to gentle gradients. The beautiful waterfalls cascading over the cliffs along the Columbia River east of Portland, Oregon, also have not had time to

Waterfall in Shell Canyon, Bighorn Mountains, Wyoming. These mountains were uplifted some sixty million years ago, glaciated during the Ice Age, and much eroded since the ice retreated.

cut their beds down to gentle slopes before their waters enter the Columbia.

Many rivers of ancient lineage antedate the mountains through which they flow. The Arkansas River in southern Colorado, for example, cut the Royal Gorge through resistant granite as the mountains were being uplifted. This scenic gash in the earth's surface is striking because of its depth and its almost vertical walls. The swift water with its rock sediments has cut deeply into the granite floor of the valley while the walls have resisted erosion in a region of only moderate precipitation.

The Colorado River is both an ancient and a young river. It is old in years since it has been draining the Rocky Mountains and Colorado Plateau country for aeons. It is young in its stage of development where it flows through the Grand Canyon with dashing rapids and many waterfalls.

If men could be like rivers they could become more youthful in their old age. But a different uplift would be required than that which rejuvenates a lazy slow-flowing river into a dashing stream with new erosive power. When the earth's crust rises beneath an already established river the cutting power of its sediment laden water increases. This is exemplified by the Columbia River gorge through the Cascade Range and by the San Juan River with its deep valley cut through the recently uplifted Colorado Plateau.

The Green River, which rises in the mountains of southern Wyoming and flows into the Colorado River in eastern Utah, provides a phenomenal manifestation of how water can saw through mountains and plateaus. From the flat Wyoming Basin the Green River enters the Uinta Mountains through beautiful Flaming Gorge canyon. Then it flows back out of the mountains only to cut back across the Uintas to emerge through the rugged canyon of Ladore and through the high Colorado Plateau to join with the Colorado River before the waters enter the Grand Canyon. The Green River was there before the Uintas were arched up some 30,000 feet. During the uplift it sawed its way down through the rocks—a perfect example of an antecedent stream.

Many rivers in the past have cut through mountain ranges instead of going around the mountains. The Susquehanna River in the Appalachians, for example, was born on the eastern slope of ancient eroded mountains that sloped toward the sea. When new uplifts occurred between the river's source and the sea, the Susquehanna cut down through the rising mountains and still flows to the sea. Some rivers in the Appalachians, while deepening their valleys near the high ridges, intercepted other rivers cutting valleys up the slopes on the other side of the mountains. The larger faster eroding river usually captured the headwaters of the smaller river and reduced its flow and eroding capability. The notch remaining in the mountain ridge, without a river flowing through its valley, became a "wind gap" as contrasted with a water gap which is a notch containing a river.

Man has always sought the easy way through the mountains. Many of the passageways permit the movement of commerce and are scenic attractions for tourists. Notable among these transmountain routes are Wilmington Notch, Smugglers Notch, Franconia Notch, and Pinkham Notch in New York and New England. Farther south in the Appalachians natural gateways have afforded passageway for travelers since settlement began on the continent. Daniel Boone on his way to Kentucky moved through Cumberland Gap. Other notable water gaps include Delaware Water Gap, the Mohawk waterway, and Newfound Gap in the Great Smokies.

In the West, where gateways over the high mountains are called passes, the routes used by early trappers and explorers now are traversed by automobile highways some of which rise to 12,000 feet or more. In the Colorado Rockies gold miners were among the first to use many passes: Loveland Pass, west of Denver; Red Mountain Pass north of the historic town of Silverton in the San Juan Mountains; and Rabbit Ears Pass which skiers use on the way to Steamboat Springs.

Donner Pass in the Sierras received its place in history when a party of settlers, bound for California, died there when winter snows put an end to their journey. The Indians, of course, knew many of the mountain passes in the West. One historic pass is Lolo Summit in the Bitterroot Mountains. The road drops down through giant red cedars into verdant valleys in western Montana where Hereford cattle graze peacefully. On September 13, 1805, the Lewis and Clark party crossed this pass on their intrepid journey to the Pacific Ocean.

Of all the ancient waters, the greatest accumulation, other than in the oceans, spread across northern North America and Eurasia during the Pleistocene epoch in the form of ice. Glaciers during the Great Ice Age, beginning about one million years ago, expanded at times until almost one-third of the earth's land surface was covered with ice sheets, some of which were thousands of feet thick. So much water was stored in the continental glaciers that the oceans were lowered several hundred feet. Of the four great ice sheets that spread southward from Canada into the United States the last one reached its maximum development about 20,000 years ago and completely melted in Canada only 6,000 years ago.

Although most people have never seen a glacier, ice sheets thousands of feet thick still occur in Antarctica and in Greenland. Various kinds of glaciers still exist, especially in Alaska and in the mountains of the western United States. A recent report by the Geological Survey estimates that more water is stored in the form of snow and ice in North America than is contained in all the lakes, rivers, ponds, and reservoirs on the continent.

The total glacier area in Alaska is approximately 17,000 square miles, and the estimated July-August streamflow from these glaciers is an almost incomprehensible 28 trillion gallons of water. Washington State has about 800 glaciers with a July-August streamflow of some 470 billion gallons. One can readily appreciate that snowmelt from this many glaciers exerts an important influence on streamflow used for hydroelectric power and farm irrigation in the dry season far beyond the mountains.

A surprising number of mountain glaciers, some of which are remnants of the great glaciers of the past, still exist in the western states. Montana has about 106 glaciers; Wyoming and California, approximately 80 each; Oregon, 38; and Colorado and Idaho, possibly 10 or 11 each. Some of these are small, but are at least 0.1 square kilometer in area. The largest glacier in the United States is Bering Glacier in Alaska, 127 miles long and 2,250 square miles in area.

Although the present mountain glaciers are as nothing compared with the great ice sheets of the Pleistocene, they tell us much of how former gigantic glaciers carved many of our mountains, changed their original landscapes, and transported their materials from mountain tops to valleys. They also made available sands, gravels, and silt that were transported great distances by water and wind.

Some of the great boulders I knew as a boy on our Nebraska farm were transported from the far north by the oldest of the continental glaciations, the Nebraskan, or the Kansan. The latter advances are known as the Illinoian and Wisconsin glaciation periods. The great moraines, or piles of rock debris, left by the melting ice at its farthest southward advance are prominent in Illinois, Indiana, and Ohio. Smaller moraines left by mountain glaciers are common in the northern Appalachians, the Adirondacks, and in the mountains of the West. The row of hills that stretches from Queens to Montauk on Long Island is a moraine which also occurs on Martha's Vineyard and Nantucket Island.

Most of the glaciers we see now in the higher mountain ranges occur in the deep hollows gouged out by larger glaciers in ancient times. Many of the great rock ampitheaters now contain cirque glaciers formed by snow and ice in the half-bowl topography below the mountain peaks. If the ice flow continues down

Glaciers in the Olympic Mountains still pluck stones from the valley sides and flow from natural amphitheaters. More than 400 inches of annual snowfall contributes to the ice streams some of which are 900 feet thick. Forward movement of Blue Glacier on Mount Olympus is about five inches per day.

slope it forms a valley glacier which grinds the rock wall and scrapes the valley bottom with rocks frozen in the ice. The awesome power of the glacier, flowing only inches each day, deepens its channel and cuts vertical walls, producing a U-shaped valley. When several glaciers emerge from their steep walled valleys they may join to form a broad ice field or piedmont glacier.

Among the striking features of glaciers are the crevasses or cracks in the ice. Some of these are hundreds of feet deep. The breakage of the ice results when the glacier moves over a high place in its channel and then bends as it goes down hill. Glaciers also crack apart when they move over precipices or when they spread out horizontally at the lower ends of wide valleys. The crevasses also channel water from the melting ice downward to the valley floor where it forms a river beneath the glacier. Thus, by observing the power of water and ice to erode the land we can understand its contribution to the shaping of mountain landscapes during millions of years in the past.

As our knowledge of past glaciation has increased, scientists now see the possibility that another Ice Age is possible. The warm period of the last 10,000 years seems to be trending toward colder global temperatures, a drier climate, especially in the great desert regions, and enlargement of the ice-bound areas in the northern polar regions. Geologists, climatologists, and paleontologists also see other signs such as the southward retreat of hardwood forests and of animals that require warmer environments. Air pollution, changes in ozone content of the atmosphere, and shifts in radiation from the earth's surface likewise influence the global temperature.

These preliminary signs of cyclic earth cooling do not mean that glaciers are going to fill our mountain canyons or that a continental ice sheet will scrape Chicago or New York off the map in our life time. If geological history repeats itself, the great glaciation could occur within 10,000 to 20,000 years. If people are living then they will be crowded southward and their space for living and growing agricultural products for food will be greatly reduced. Men, however, will be able to travel over and explore the great ice sheet just as they explore the ice cap in Greenland today. And the edge of the ice sheet will virtually border on the coniferous forests in the mountains and practically touch the grasslands of the prairies and plains. Man's ingenuity undoubtedly will enable the race to cope with the new conditions. It might be a fascinating time in which to live.

Most of the great rivers of America rise in the mountains. The Mississippi seems to be an exception since its source is in the low lands of Minnesota. But much of its water comes from the Missouri, the Platte, the Arkansas and other rivers born in the Rocky Mountains and from the Ohio and other rivers that run their courses from the Appalachians to the Gulf of Mexico. The great rivers all run to the sea but some lesser streams drain into the Great Basin where their waters sink into the parched sands of the desert.

The wasting of mountains by wind and water is especially apparent in desert country where rivers have no outlet to the sea. Anyone who has visited the Great Basin or has vacationed in Death Valley must have noticed how the land-locked valleys are filled with rock debris and silt from the surrounding hills. Even the foothills are covered in many places and the boulder strewn slopes attest to the power of water as it comes down in occasional floods.

This universal erosion is more pronounced in regions of low rainfall and high wind movement. Plant life on the lower slopes of desert mountains is sparse and has minimum effect on soil stabilization. Frost also attacks the bare rocks causing debris to fall from cliff tops as in the Grand Canyon. As soils and rocks are loosened by weathering and as soluble salts are leached from the land their downward movement is powered by gravity. And the rills, creeks, and rivers are always there as avenues for transport to lower elevations.

Erosion is less apparent to the casual observer in mountains where plant life is abundant. In the Rockies, even though hard rock layers are almost constantly in view, the forested slopes and grassy meadows release their water more slowly and the debris load is much less than in desert mountain streams. Where trees, shrubs, and other plant life grow luxuriantly, as in the Appalachians, the streams usually carry sparkling pure water indicating the near absence of accelerated erosion. Talus slopes, alluvial fans, and landslides are rare.

Rivers, from their sources to their demise in the ocean, have

Millions of fishermen find their greatest sport in clear dashing mountain streams. The author here is fishing for trout in the Cascades in northern Washington.

many characters and many moods. The rivers of the Coast Ranges along the Pacific slope are in mountains all the way from their sources until they flow into the ocean. Their waters are turbulent and fast moving during most of their seaward journey. Rivers that flow into the midlands of America forsake their narrow channels as they leave the foothills and become deep placid drainage ways or wide meandering ditches, familiar to all who drive the nation's low level highways.

One can never really know a river until he has seen its source and has followed it to its discharge into the sea. I am thinking now of the mighty Columbia River. For several years I could look down from our front yard in Portland, Oregon, and see the sail boats, water skiers, and ocean going vessels plying the mile wide river, just above its junction with the Willamette River which collects its waters from the Cascades and the east slope of the Coast Range. With my boating friends I fished for sturgeon in 90 foot deep water west of Vancouver, Washington and crossed the bar at Astoria to troll for salmon on the heaving water of the Pacific Ocean.

Other exploration trips took me to the headwaters of the Clackamas River where it flows down from the Cascades south of Mount Hood to join the Willamette River. Then I followed the Columbia tributaries, the Deschutes, the John Day, the Yakima, the Wenatchee, and the Okanogan Rivers to their beginnings in Oregon and Washington. Before I saw all these I fished in the Snake River near its source in the Yellowstone country in northwestern Wyoming and nearly drowned while demonstrating my swimming prowess to my wife on our honeymoon.

I hunted deer and bear in the awesome depths of Hells Canyon south of Lewiston, Idaho. And before the Army Engineers stopped the waters of the Columbia, I photographed the Indians pulling up salmon to their perilous perches on board platforms suspended above the thundering water at Celilo Falls. Finally, I made the journey to Columbia Lake in British Columbia to see the headwaters between the Canadian Rockies and Selkirk Mountains where the river begins.

From Columbia Lake the river flows northwest for 218 miles and then south for 280 miles to the United States-Canadian border. The road on the northward journey winds through vivid spruce-fir forests on the lower slopes of the magnificent Canadian Rockies. I have not been there for several years, but when I made the journey I hoped there would be gasoline at the filling station at Boat Encampment where the river makes its northernmost turn. There was gasoline. And there was no anxiety about returning to Revelstoke where the paved highway from Calgary and Banff to Kamloops and Vancouver entices one to further travel.

The basin of the Columbia River includes parts of five physiographic provinces and drains an area of 259,000 square miles. From rugged mountain terrain of sedimentary and metamorphic rocks it flows through volcanic igneous rocks in its lower reaches. Most of the mountainous area of the basin is characterized by stands of Douglas fir, hemlock, and pine. Where the

Streams in the rugged Wallowa Mountains in eastern Oregon contribute their waters to the Snake River which in turn contributes to the mighty Columbia. So many dams have been built on the Columbia that much of its water runs through turbines which generate electric power for the Northwest. Without mountains this energy source would not exist.

river traverses the Columbia Plateau the terrain is underlaid by lava and the region is semi-arid sagebrush desert. When one realizes that commercial shipping goes as far inland as Pasco, Washington, that three and one half million people use about three billion gallons of ground and surface water each day, and that the river is one of the world's greatest sources of hydroelectric power, it is easy to appreciate the contribution that this network of mountain rivers makes to this Northwestern land and to its people.

Some rivers have been immortalized by great writers in the past. Who has not heard of, or read, *The Adventures of Tom Sawyer* on the Mississippi? Another story that has enthralled me since childhood is the account of Rip Van Winkle's visit to the mountain glen with the ninepins and the keg of wondrous brew. Washington Irving began the tale, "Whoever has made a voyage up the Hudson must remember the Kaaterskill Mountains. They are a dis-membered branch of the great Appalachian family, and are seen away to the west of the river, swelling up to a noble height and lording it over the surrounding country."

Who could resist such an invitation to visit the Catskill Mountains or the Adirondacks where the Hudson begins its flow out of Lake Tear of the Clouds in Essex County, New York? The Hudson is civilized in its lower reaches below Troy, in spite of its picturesque rocks in the Highlands and the Palisades above upper New York Bay. But the river is surrounded with history, beginning with its discovery in 1524 by Giovanni da Verrazano, an Italian sailor, and its exploration by Henry Hudson in 1609. Indians were the first inhabitants in the basin. Dutch colonists were among the first settlers and around 1624 they established New Amsterdam on Manhattan Island. Now the great river provides water for millions of people and serves agriculture, industry, and commerce in one of the heaviest populated regions in the nation.

There are other great and small rivers that begin in the mountains and flow to the sea that are worthy of exploration and understanding. The Potomac, for example, rises in the western Allegheny Mountains and derives its waters from the Appalachian Plateau, Valley and Ridge, Blue Ridge, Piedmont, and Coastal Plain provinces. In the upper reaches of the basin the land is rugged and mountainous with exposures of folded sedimentary rocks including limestone, sandstone, and shale. Metamorphic and igneous rocks predominate from the Blue Ridge to Washington where the river meets the tide. Below Mount Vernon it becomes an estuary and then enters Chesapeake Bay about seventy-three miles from the Atlantic Ocean.

The Delaware is another river of great human value—6½ million people use its water which is an important source of supply for New York City, Easton and Philadelphia, Pennsylvania, and Phillipsburg and Trenton, New Jersey. Like the Hudson, the Delaware River and Delaware Bay were explored by Henry Hudson. The river begins on the western slopes of the Catskill Mountains, travels through wooded hills, continues through the Kittatinny Mountains, and passes through the famous gorge known as the Delaware Water Gap. It flows past Trenton, New Jersey and Wilmington, Delaware into Delaware Bay.

There are innumerable places in America where one may see the handiwork of mountain rivers. To see and explore even one of the great rivers and its tributaries over a period of years makes one humble and develops appreciation of the gifts of mountains and water in this land we love. Understanding of the strange working of mountain rivers does not come easily. A little knowledge of geological sequences and events is helpful. An awareness of the sculpturing effects of wind, frost, snow, ice, water, and gravity impels one to look more closely at the evidence that water is forever sawing and disintegrating the face of the land. It is also the life supporting fluid for plants and all the moving creatures of earth.

One of the great mysteries of earth that has puzzled man since the time of the ancients is the origin and nature of ground water. In mythological times it was believed that Neptune, God of the Sea, poured water into a dragonlike creature which then squirted rain over land and sea. But the ancients left no record of great knowledge of underground water which is such an essential part of man's environment. Today we know that some of the water which falls as rain seeps into the ground and saturates sand, gravel, and rocks. It fills pores and cracks until it reaches an impervious layer such as clay, shale, or granite. The top of the saturated portion is called the water table.

The presence of ground water shows itself both in subtle and in astounding ways. The surface of many rivers and lakes often is the visible level of the water table as is the height of water in a well. Water flowing from a spring indicates a water table above the level of the spring. And most striking of all is the explosive eruption of geysers, the boiling of hot springs, and the emergence of full grown rivers from lava rocks in volcanic country or from caves in limestone regions.

Underground waters give us some of the most attractive features of mountains. The great caves with their labyrinthian tunnels and cathedrallike rooms created by solvent action of water. Minerals redeposited by dripping water decorate the floors and ceilings of caves with stalagmites and stalactites of fantastic shapes and colors. Some caves, where underground temperatures are low, have ice-coated walls and floors. There is even a frozen underground river in Crystal Falls Cave, northwest of St. Anthony, Idaho.

Underground water flows through rocks, especially those that are jointed or are filled with natural pore spaces. Water travels through gravel faster than it flows through sand or silt. Contrary to popular opinion most underground water does not flow in subterranean streams or rivers. Instead it moves in aquifers which are layers of gravel, sandstone, or other porous rocks.

Aquifers that outcrop on mountainsides and then dip below the land surface sometimes carry water for hundreds of miles. One of the greatest of these collects water from the eastern edge of the Rockies and the Black Hills and transports it between layers of impervious rocks far to the east in the Dakotas, Nebraska, and Kansas where it is pumped from wells for domestic use. How long the water takes to travel from the Rockies to eastern Nebraska, no one knows. In silt or fine sand it may travel only a few inches a day.

Artesian wells also are a gift of the mountains. When water enters an aquifer overlaid by an impervious layer of rock, it builds up pressure the deeper and farther away it flows from the mountains. When a well is drilled into the aquifer, even hundreds of miles from the mountains, the water may spout into the air and run constantly for long periods of time since the source of supply is independent of local rainfall. Rocks that yield artesian water have been found at depths of more than 6,000 feet.

The wildlife communities associated with mountain aquatic habitats exhibit an extraordinary variety. Environmental conditions are rigid at the edges of alpine snowfields and in the rills that coalesce to form little waters, and then streams, and finally rivers. The waters are icy cold, snow and ice frequently cover plants and aquatic animals, and the growing season is short. Downhill, the river abandons its open pools of smooth water with sandy bottoms, tumbles over boulders, thunders through rocky canyons, and then becomes a series of riffles and pools with backwaters and eddies where fish find it easier to cope with the force of the current.

Almost every yard, and certainly every mile, in the downhill course of a river, offers a different habitat for the aquatic creatures in the food web that extends beyond the waters of the river itself. Sandy bottoms provide spawning areas for trout and salmon. Riffles aerate the water, and pools slow the current so moss and algae can grow. Stones provide attachment places and shelter for numerous stream animals such as caddisflies, May-

Snowmelt in the mountains supplies water for many lakes. Beaver dams maintain the water level throughout the seasons in small lakes and ponds. This provides a stable habitat for trout.

fly, nymphs, and planarians. These the trout and the dipper use for food while the kingfisher dives from his streamside tree for the unwary trout.

As the river receives water from smaller watercourses and grows in size it becomes torrential in the higher elevations and cascades over waterfalls. The spray from falls moistens the surrounding rocks and provides a microhabitat for mosses and animals that live on wet stones. Farther down there is a gradation from river bank to forest groves or grassy meadows. The transition between aquatic habitat and forest grove usually is abrupt, but the edge effect is congenial to many birds and animals. The beaver, for example, builds his house and dam in the stream but his food supply of aspen and willows grows on land. The raccoon wanders far and wide between the river and his woodland territory. The deer browses on the mountainside but comes down to the river to drink.

The mineral nutrients in the river also vary according to the soil and rock materials from which mountain waters flow. Rivers flowing over limestone, gypsum, and marble, dissolve greater quantities of salts than rivers flowing over granite and lava. The amount of dissolved minerals also varies between the source and the mouth of the river. The Dolores River in southwestern Colorado provides an extreme example. It tumbles out of the mountains as a freshwater stream, passes over an ancient salt dome, and flows toward the Colorado River as salty as the ocean. Scientists estimate that it adds 200,000 tons of salt each year to the Colorado River system. The effect on animal life in the river and on water quality for irrigation and municipal use downstream has even caused the Mexican government to demand remedial measures which would improve Colorado River water quality at the international border.

The Rogue River in southwestern Oregon is another kind of river. It is born in the icy springs in the lava rocks of the Cascades at the foot of Mount Mazama. It spends its youth wandering through forested mountains and cascading into the Rogue Valley near the cities of Medford and Grants Pass. Then it rampages through Rogue Canyon to the settlement of Agness where it flows more peacefully to the Pacific Ocean at Gold Beach. It gets help from Elk Creek, Bear Creek, Applegate River, Illinois River, and a multitude of lesser tributaries. The combined strength of these waters springs from the mountains and their proximity to the sea provides a choice route for fish migration.

Spring chinook salmon enter the river in April and May. The fall chinook migration follows soon after and extends into December. Formerly, large runs of coho reached the headwaters of the Rogue and its tributaries but dams and other impacts of civilization have reduced their numbers. Summer steelhead still flock to the river in August and September. The winter steelhead run lasts from November to May with most spawning in the gravel beds in March and April. Other anadromous fish include sea-run cutthroat trout, shad, white and green sturgeon, and an occasional striped bass. Brook trout are found in some of the higher elevation tributaries while native cutthroat and rainbow trout occur in the lower elevation tributaries. Brown trout are seen in one short stretch of the North Fork of the Rogue.

Truly this is a sportsman's river where the fish are nurtured by the physical and biological substances of mountains. No wonder that Zane Gray was enchanted by its beauty and helped to make the Rogue famous by his writing.

Mountain rivers are the most visible reminders of the eternal battle between water and land. But the hydrological cycle includes more than rain, erosion, and return of the rocks to the sea. Without water's life-giving substance plants and animals would not exist. And there would be no glaciers, no snow on the peaks, and no clouds to quench the thirst of the land. Nor would the sculptured landscapes have the unique majesty and beauty we now see in the mountains.

Aspens are widely distributed across the northern part of the continent. They are among the first trees to appear when fire destroys the mountain forests. In this aspen forest young spruce trees are growing and ultimately will shade and replace the aspens.

5

Mountain Forests

IN THE EARLY YEARS of the United States luxuriant primeval forests clothed nearly one-third of the nation. Stretching from the Atlantic Ocean to west of the Mississippi and southward into central Texas an almost unbroken canopy of trees covered the land. Extensions of this forest followed the rivers and streams still further westward through the prairies. In northern Minnesota, Michigan, and New England, the forest consisted primarily of conifers. This northern evergreen forest also grew on the slopes of the higher mountains from Maine to Georgia.

At lower elevations along the coast from Massachusetts to New Jersey, and inland on the Piedmont east of the Appalachians and south to Alabama, the great central forest of deciduous trees covered the land. Deciduous trees also dominated the midlands from Ohio to Tennessee and from Wisconsin to Arkansas and Texas. Depending on locality and site conditions the forests consisted of oak, hickory, maple, beech, poplar, and other species in a remarkable variety of associations. Bordering the deciduous forests were the piny woods on the coastal plain from New Jersey to eastern Texas. In this border forest were short-leaf pine, long-leaf pine, and loblolly pine on the drier sites. Tupelo, magnolia, sweetgum, and cypress grew in the swamps and moist soils.

West of the prairie, the Rocky Mountains were forested with stands of spruce and fir at high elevations. Ponderosa pine, aspen,

and lodgepole pine grew at medium elevations. Oaks characterized many of the foothills in the Southern Rocky Mountains.
The Rocky Mountains from Canada to New Mexico, and the
high mountains in the Great Basin supported luxuriant coniferous forests on their slopes below the Arctic alpine zone.

Beyond the desert, in the Sierra Nevada, the Cascades, and
the Coast Ranges were some of the most magnificent forests of
the world. Here were the giant sequoias in the Sierra, the redwoods in the fog shrouded slopes inland from the sea, and the
giant Douglas firs, Sitka spruces, western hemlocks, and western
red cedars that seemed to reach to the sky over cathedrallike
aisles carpeted with ferns, salmon berries, and other shrubby
undergrowth.

In all of these major forest divisions, from the Atlantic to the
shores of the Pacific, the tree species appeared in different mixtures, depending on local habitats. Topography and altitude
exerted the greatest influences. Wherever mountains rise within
the great forest divisions, tree zones appear as great belts that
succeed one another as the elevation increases from foothills to
timberline. Within these belts, wind, light, temperature, precipitation, and soil types determine the kinds of local communities
of trees.

If you want to see the scenic beauty and the ecological group-
ing of forest types, a good place to begin is in Appalachia or in
the Adirondacks. You will have to climb a mountain, preferably
on foot. If you want to see the variations in life zones from north
to south you should climb several mountains. Katahdin in
Maine is the difficult one, not accessible by automobile, but
hardly surpassed in grandeur by any other mountain in the East.
Mount Colden, Mount Marcy, or Algonquin in the Adirondacks
have easy trails that lead upward from deciduous forests in the
low elevations to dwarfed black spruce trees near their summits.
The Mount Washington toll road in the White Mountains allows easy ascent through all the forest zones to the treeless alpine
slopes in the Presidential Range.

These are only the beginning. To broaden your concept and
understanding of elevational zones you must see the magnificence of the flora in the Applachian mixed hardwoods in the
coves and on the slopes of the central Appalachian Mountains
and the adjacent portions of the Allegheny and Cumberland

Plateaus. Northern red oak and yellow poplar are intermingled with white oak, American beech, black cherry, cucumbertree, hickories, yellow buckeye, black locust, and more than two dozen other trees. The species proportions vary with aspect, elevation, and latitude.

Do not expect the Appalachians to be the great wilderness encountered by the white men when it was still sacred to the red men. Some of the forests have been logged five times since the Colonial expansion from 1620 to 1776 began exploitation of the land for permanent settlements. As the frontier moved westward increasing use of forests for commercial purposes denuded the woods on both lowlands and mountains. The forest materials needed for growth of the new civilization seemingly were endless. Forest products went into the building of homes, stockades, ship masts, charcoal for smelting iron, cordwood for fuel, hemlock bark for tanning leather, and potash from hardwood ashes for the glass and textile industries.

Repeated logging of forests has resulted in widespread plant community replacement, a process called ecological succession. The successions which originally began on exposed rock surfaces in the mountains progressed through many stages until the final mature communities, or climax stages, were reached. This process, known as primary succession, required thousands of years, especially in mountains denuded by glaciers during the Ice Age. When a forest is logged, however, the land is not made barren of all vegetation and the recovery process, called secondary succession, is comparatively rapid since many of the plants and animals are already present when the renewal begins. Forests which have been logged commonly produce another stand of trees within 50 to 100 years.

The new stand of trees produced by secondary succession frequently differs greatly from the original or virgin forest the pioneers saw when they explored the mountains. The physical environment sometimes is greatly modified by erosion following logging, strip mining, and by fire, whether man-caused or natural. Environmental disturbance also changes the proportions of mammals, birds, insects, and plant species originally present in the virgin forest. This results in a different successional direction and the establishment of biotic communities with different habitat relationships. If the dominant trees are removed, or are killed by disease, as were the American chestnuts in the early part of the century, the ultimate stable community may never be the same as the original climax. Whether this is desirable or not from an economic standpoint, aberrations in succession do account for the remarkable variety of plant communities we encounter in the mountains.

Much diversity in animal life arises from the presence of innumerable succession stages in the mountains. Grasses and low growing herbs, for example, frequently invade newly denuded forest areas. In this early stage of stabilization of the land the invertebrate population may consist of species of ants, grasshoppers, spiders, snails, and worms not common to later stages of succession. The predators of these animals also will be different. Sparrows and meadowlarks, mice and shrews, and other "grassland" animals will be a part of the food chain. If the grassland changes to pine forest, wrens, chickadees, woodpeckers, and flycatchers may be part of the bird population. And if the forest eventually becomes an oak-hickory climax one may expect to find squirrels, warblers, blue jays, wild turkeys, deer, opossums, and other animals adapted to this type of environment where food, shelter, and local climate make their lives possible.

Inspection of the strata or layers in a biotic community such as an oak-hickory forest, can reveal why many organisms are where they are and what they do. Some birds, the thrushes for example, seek their food near the ground while others, such as the blackburnian warbler, seek the levels of the higher trees. The periodical cicadas use different levels during their life cycles. As larvae they spend as much as seventeen years deep in the soil. When they emerge as winged adults they fill the late summer air with their rasping songs high in the trees. They lay their eggs in the twigs of more than seventy species of trees; the most susceptible are the oaks, hickory, honeylocust, gum, walnut, and ash. Their presence in trees and shrubs attracts numerous birds.

Secondary successions, particularly those induced by cataclysmic disturbances such as fire, avalanches, or clear-cutting which removes all trees, exhibit a variety of developmental stages that contrast markedly with adjoining communities which have not been disturbed or which are in different stages of succession. The transition zones along the boundaries of these contrasting communities often contain plant species from both habitats and hence an increased variety of food for animals. Many birds, for example, nest in shrubs and trees in the forest and search for insects in the border zones. Rabbits that feed in open spaces also find shelter in bramble patches that occur in the "edge" zone next to the forest. With close observation it is possible to see many other general relationships between species and numbers of plants and animals which are influenced by ecological succession.

White pine (*Pinus strobus*), one of the noblest of trees, was the monarch in its domain from Maine to Pennsylvania and south on the Appalachians to Georgia. Its pyramidal outline accentuated by whorls of branches heaped tier on tier to heights of 150 feet or more, and clothed with soft fine needles, five to a bundle, astounded the pioneers. Henry David Thoreau wrote of the beauty of the white pine in Maine and thought of how it would be brought low by the axe and sold to the New England Friction Match Company. Contracts were let by England for great mast sticks for ships of the navy and of commerce. Some of the tree trunks were straight and free of branches for eighty feet above the ground—enough for five 16-foot saw logs.

For lumber, the wood of the white pine was unrivaled by any other tree. The grain was straight; the wood was soft, light, and strong; it stained well when used for paneling in homes, and it milled easily for doors, window-sashes, and utility furniture. Its long-lasting qualities made it a choice material for covered bridges and shingles for house roofs.

With all this exploitation the last of the great virgin trees to be felled for industry, mainly between 1900 and 1915, were the white pines in the coves of the southern Appalachians. Now, second growth of white pine from Maine to North Carolina, is increasing the production of board feet of lumber. But the magnificence of the new trees will never be the same. Nor will the composition of the forests in which they grew.

Professional forest management now tends toward propagation of single species and even-aged stands. This causes the elimination of secondary tree species which once contributed to the variety and charm of the mountain forests. Release cutting of less desirable species is also done to hasten conversion to types with greater commercial potential. Even the valuable species are harvested on a cutting cycle that never allows the longer lived trees to attain the magnificent size once attained in the aboriginal forests. Only in some of the wilderness areas and national parks may we or future generations hope to see woodlands redeemed through natural growth to a semblance of their former glory. The process may take from one to three hundred years.

Even with non-interference by man, many forests may never

Virgin stands of Douglas fir in the Northwest are rapidly vanishing before the loggers' chain saws. The second growth forests will never be allowed to grow to this size because of the insatiable demand for lumber, plywood, paper, and other forest products.

return to their former composition. A recent study by Halkard E. Mackey, Jr. and Neal Sivec provides a record of how a former oak-chestnut forest in the Allegheny Mountains of western Pennsylvania is regenerating in a pattern not typical of previously studied chestnut sites. The forest, apparently once dominated by chestnut, oaks, yellow poplar, basswood, and pitch pine, now is occupied by black cherry (*Prunus serotina*), red maple (*Acer rubrum*), sugar maple (*A. saccharum*), black oak (*Quercus velutina*), black birch (*Betula lenta*), and sour gum (*Nyssa sylvatica*). The trees now involved in the new stand undoubtedly were present in the original forest. But the proportions of trees that originally were important is much different now and will continue to be different in the future.

The person who wishes to see trees in great variety in America should visit the Great Smoky Mountains National Park. Here, because of elevation, climate, and geological influences in the past, are trees of subtropical affinities, trees of the mid continent, and trees that have survived after their ancestors were pushed southward during the Ice Age. The cold climate trees are on top of the mountains; the warmth-loving trees are on the slopes and in the valleys and coves. The pleasure of learning to identify at least a few of these trees, while rambling up and down the slopes in this part of the Appalachians, will also evoke an interest in how forest associations developed and how they are related to environment.

In the Great Smokies, especially in the coves, you will find the great melting pot of northern and southern trees. Here, growing side by side with beeches and sugar maples are Carolina silverbell (*Halesia carolina*), magnolias, and yellowwood (*Cladastris lutea*). From Gatlinburg, Tennessee, you can see much of this forest variation by driving the highway through Great Smoky Mountain National Park. The road leads nearly to the top of Clingmans Dome but you will have to leave your car to hike to the top of Mount Le Conte (6,593 feet). Now you will be in the red spruce (*Picea rubens*) and Fraser fir (*Abies fraseri*) forest that resembles the black spruce (*Picea mariana*) and balsam fir (*Abies balsamea*) forests of the northern Appalachians. On these southern mountains, especially in the transition zones below the summits, are multitudes of attractions for the naturalist, ranging from rhododendrons, bears, strange salamanders, famil-

iar and unfamiliar shrubs, and a Pandora's box of showy wildflowers on the forest floor.

Of all the trees in the great central and eastern forests the oaks
are among the most valuable lumber trees, the best known, and
well loved. One species or another occurs in most of the mountain ranges of the United States. What one of us in our childhood, with the exception of those born in great cities, has not
had a favorite oak tree with its bird and animal inhabitants, its
welcome summer shade, and its distinctive acorns? And how
many of our fathers or grandfathers warmed themselves once
while gathering oak logs for fuel, and again by the winter fire
of oak in the fireplace or in the glowing stove in the family par-
lor? The usefulness and enjoyment was there whatever the species of oak, of which there were as many as sixty in some states.
As a group the oaks are easily distinguished by their acorns.
In general the species belong to two groups, the white oaks and
the red oaks. The leaves of oaks in each of these groups vary.
Some are lobed; others are coarsely toothed without lobes; and
a few have smooth margins, are evergreen, or remain on the
twigs until the following spring. Leaves of the white oak group
have rounded lobes, if lobes are present, and there are no bristles
on the tips. The acorns are sweet to the taste and ripen in one
growing season. Leaves of the red oak group have pointed or
angular lobes, if present, and a bristle at the tip of each lobe. The
acorns are usually bitter and ripen on twigs formed in the previous year. The bark of trees in the white oak group is usually
gray. In the red oak group the bark is usually very dark.
Some of the red oaks are called willow oaks since their leaves
resemble willow leaves. Most of these are common on the coastal
plain but shingle oak (*Quercus imbricaria*) grows on the lower
mountain slopes from Pennsylvania and Virginia west to Arkansas. The wavy margined leaves of shingle oak are shiny
above and rusty-hairy below at maturity. The chestnut oak (*Q.
prinus*) and the chinkapin oak (*Q. muhlenbergii*), both of wide
distribution, are hillside trees commonly found with white oaks
(*Q. alba*) and northern red oaks (*Q. rubra*), shagbark hickories
(*Carya ovata*) and sugar maples (*Acer saccharum*).
As timber trees, the oaks include some of the most valuable
and important trees in America. Many grow to heights of sev-

enty to eighty feet and a few, such as the northern red oak and the black oak attain heights of 100 feet or more and diameters of three to four feet. Oaks of smaller size, such as the blackjack oak (*Q. marilandica*) which grows on dry stony mountain slopes, are used mainly for fuel and charcoal. As a group, however, all oaks are useful for ground cover and soil protection. In fall and winter their acorns furnish food for songbirds, game birds, deer, and small mammals. Their importance to wildlife and the forest environment make their identification of value to everyone interested in nature.

In the Appalachian hardwoods are many trees of lesser stature. They add much to the variety and beauty of the forest. From early spring to late autumn they influence the aspect of the forest through colorful displays of flowers, differing leaf arrangements, and attractive fruits that bring new combinations of birds and mammals to glean the autumn harvest. Looking at the trees instead of the forest, particularly the understory species that grow beneath the dominant forest trees, the observer can note a never ending source of unique and interesting botanical features.

Some of the lesser trees are spectacular at flowering time. The redbud (*Cercis canadensis*) bursting its rose-purple flowers before the leaves appear, lends charm to the woods already made ethereal by the warm sunlight filtering through still leafless branches of larger trees. Later, its shining heart-shaped leaves distinguish it from all its neighbors.

The flowering dogwood (*Cornus florida*) brings exclamations of delight when its florist's arrangement of white blooms, tier on tier, accentuate the darkness of the woods or highlight the pastures and forest clearings. It is not choosy about the soil on which it grows and it adapts to streamsides as well as to high mountain slopes. The flowers themselves are inconspicuous; it is the four petallike bracts, white, streaked with green, and notched at the tip, that show their brilliance in the spring sunshine. But the tree is not done with handsomeness until its brilliant shiny scarlet fruit clusters add yet another design to the autumn scene.

Another tree that adds to the witchery of the autumn woods is the witch hazel (*Hamamelis virginiana*). Its common name comes from its legendary magical properties and especially from its choice as the favorite wood for forked branches used by water-dowsers. The tree appeals to me because of its botanical strangeness. Not until the broad wavy-margined leaves drop in fall do the flowers appear with their yellow stringy twisted petals, half an inch long, inconspicuous on the brownish twigs. But stranger still is the rattle of the hard black seeds against the fallen leaves on the forest floor. These are ejected forcefully from the fruits ripened from flowers of the previous year. These little projectiles are expelled with such violence they travel up to twenty feet or more and thus contribute to propagation of new witch hazels beyond the mother tree.

One tree you will not see in the Appalachians is the American chestnut, once the mighty monarch of the forests. Some specimens rivaled the Douglas firs and Sitka spruces of the West with trunk diameters of twenty-five to thirty feet. The spreading crowns, however, did not attain the great heights of the western conifers. Still, this was the dominant tree among the oaks and all of its other deciduous associates. For lumber, shingles, and other economic uses its wood had no equal; and its nuts, enclosed in spiny burs, were palatable for man and wildlife.

The chestnuts died with sickening rapidity. The blight, apparently introduced from China about 1904, resisted all attempts at control. The spores of the blight, carried by the wind, infected the trees from Maine to Alabama and from Ohio to Mississippi. When I saw them in 1925 and in later years the trunks stripped of their bark stood like ghosts along the forest trails. Although sprouts appear from some of the stumps there seems to be little hope for their survival.

As late as the mid-1930's, I used to surprise my dendrology students at Marquette University with fresh specimens of the American chestnut, including the crisp green leaves with prominent veins running to the marginal teeth, the male catkins with their yellow flowers, and the prickly burs that split open in autumn to reveal the palatable chestnuts within their shells. I never told anyone where these trees grew. Since they were nearly a thousand miles from their native range I nurtured the hope they would escape the blight. In June, 1973, after an absence of twenty years, I visited them again on the farm of Robert Trail, near Julian, Nebraska. They were still standing in the dooryard —bleached, whitened skeletons, still higher than the elms, oaks, and hickories in the nearby forest. The blight had found them out.

Even though the chestnuts are gone, the Southern Appalachians still have trees of notable size. Sugar maples, yellow-poplars, mountain silverbells, and yellow buckeyes with breast high diameters of four to six feet are not uncommon. Other trees that make record growth are basswood, American beech, eastern hemlock, cucumbertree, and northern red oak. The environment is so congenial that more than 100 kinds of trees make up the silva of the area.

The greatest variety is found in sheltered valleys at moderate altitudes. Also common are black cherry, red maple, pin cherry, and American holly. On the high ridges, red spruce and Fraser fir reach above the 6,000 foot level where they represent southern extensions of the great coniferous forests of the far northern mountains. Associated with them are American mountain-ash trees which you can also see on the slopes of Mt. Washington in New Hampshire and still farther north in Quebec and Newfoundland.

In the forests of the Ozark and Ouachita Mountains the roads meander through pleasant valleys with clear streams born in limestone strata in the north or flowing from granite exposures in the south. Rising steeply above the valleys are green hills and mountains, once covered with primeval forests of oak and hickory, interspersed with short-leaf pine, but now mostly second growth oak intermingled with many other tree species. The forest cover is a western extension of the deciduous forests of the central eastern United States and includes many species that grow in the southern Appalachians.

The Ozark Highlands of Missouri, covering approximately one-third of the central, southern and southeastern part of the state, now is an area of hills and broad smooth valleys, worn down from a geologically old plateau. Although most of the localities are less than 1,400 feet in elevation, steep canyons and rocky knobs occur in some places. Most of the original timber has been harvested but the slopes still support considerable areas of oaks, butternut, hickories, and persimmons. Near the south-

Second growth timber now covers much of the Ozark Mountain area in Arkansas. The original timber was cut many years ago. The new growth includes pine, gum, oak, hickory, and a variety of other hardwoods.

ern border, red cedars, cherries, black gum (*Nyssa sylvatica*), walnuts, and sugar maples also add variety to the forests.

In Arkansas the Ozark Highland, a dissected plateau, rises northward from the Arkansas River to the summits of the Boston Mountains. The country continues rugged to the north as the altitude decreases toward the Ozark Highland of Missouri. In this region in Arkansas the black bear once was common. Other characteristic mammals are raccoons, red foxes, opossums, flying squirrels, chipmunks, and striped skunks. They live in forests with tree communities similar to those in the Missouri Highlands.

South of the Arkansas River the Ouachita Mountains are marked by long parallel ridges, some of which reach 2,800 feet, with broad valleys between the ridges. Hard sandstone, slate, and shales form the ridges whereas limestone outcrops compose much of the horizontal sedimentary strata in the Ozark region north of the Boston Mountains. Hard maple (*Acer saccharum*) appears on these limestone outcrops whereas the sandstone ridges of the Ouachita Mountains offer more favorable sites for oaks and hickories on the north slopes and for shortleaf pine (*Pinus echinata*) with the hardwoods on south slopes.

The green wooded hills of Arkansas offer an assortment of recreational resources, many of which are related to the forests that grow there. Picnicking and camping spots are numerous on the Ozark and the Ouachita National Forests. From these camp grounds, excursions may be taken for study of trees, wildflowers, mammals, birds, and geological formations in a varied forest environment. Some areas, such as the Blowout Mountain near Mount Ida, west of Hot Springs, have been set aside for nature students because the spots have been relatively undisturbed by man in the past.

In Hot Springs National Park, where forty-seven springs flow at a constant temperature of 143 degrees, nature trails lead past trees that are named on labels. Farther west, Talimena Skyline Drive traverses the ridge of the mountains with scenic views of the forested country, and from which hiking trails permit contact with a remarkable variety of plant life. The scenic drive winds for 55 miles along the crests of Rich Mountain and Winding Stair Mountain between Mena, Arkansas and Talihina, Oklahoma.

Forest products in these mountains have had their effect on villages and the hill people. Clapboard houses, some plain and some with gingerbread trimmings, are vintages of former times. The rustic life in Ma and Pa Kettle type houses is still real, although maybe a bit exaggerated by Lum and Abner of former radio fame; their store and museum still stands at the edge of the woods at Pine Ridge, Arkansas.

To the Ozarkians, trees have been a part of their lives for

generations. They have long known which timbers are strong and sturdy for building their cabin homes; which oak is best for tables; which hickories to use for handles and implements. At harvest time they collect the fruits of the walnut, hazel, hickory, and pecan. The meats of these nuts are spice for candies, cakes, cookies, and brown bread. When the sassafras changes to orange, the wahoo flaunts its red leaves, the oak turns russet, copper, and brown amid the flame of hard maples, it is time to collect pine cones for Christmas decorations. When the cold of winter comes there is ash for quick fires and oak and hickory for long-burning fires. Then cedar and pine are added for aroma in the fireplace.

The forest types in the Ozark-Ouachita Mountains Region vary with slope, aspect, and elevation. On north facing slopes and ravines, white, red, and black oaks are dominant. With these grow white, northern red, bur, and chinkapin oaks. The list of species also includes shagbark, mockernut, and bitternut hickories; sugar and red maples; white ash, black gum, black walnut, beech, and cherry. On drier south facing slopes and other less favorable sites, post oak, blackjack oak, and black hickory (*Carya texana*) are common associates. With these are winged elm, persimmon, eastern red cedar, shortleaf pine, red haws, and blackjack oaks. In mountain stream valleys where the timber has not been cleared for farms are various combinations of silver maple (*A. saccharinum*), river birch (*Betula nigra*), and slippery elm.

Sassafras (*Sassafras albidum*) is one tree that stirs my childhood memories. Because of folklore my grandmother made sassafras tea each spring and gave it to me to drink. "It thins the winter blood and wards off sickness," she said. The advertisers' antibiotics and little handy-dandy, super-duper sickness pills, available at drugstores, have long since replaced the tea. But I still remember the sweet flavoring of wild honey in the tea and the pleasant aromatic smell of the root bark she kept in the kitchen pantry. Years later, when I first saw the tree itself in the Indiana dunes, it seemed like an old friend, but with new idiosyncrasies in addition to its fancied cure-all properties.

Sassafras flowers appear in small greenish yellow clusters on the bare green or reddish brown twigs. Quickly the leaves follow and they are as varied in shape as mulberry leaves. Some are mitten-shaped, others are three-lobed, and still others are ovate with unlobed margins. In autumn the red orange, vermillion, or lemon yellow leaves add splendor to the southern forests. In another colorful gesture the spicy blue or black fruits appear in scarlet cups on long scarlet stalks. In favorable sites the tree may attain a height of 80 feet and a breast high diameter of 4 feet.

The papaw (*Asimina triloba*) is a strange tree of the understory on moist northern slopes and in protected coves. Its broad leaves, eight to ten inches long, the two-inch wide flowers, green at first and then turning brown or purple, and the banana-like fruits which become edible in late autumn give these small trees an aura of tropical attraction. In November the skin of the fruit wrinkles and dries to greenish yellow or dark brown and the custardlike interior becomes soft and palatable. Opossums, raccoons, and squirrels eat them avidly.

One of the big trees which may attain a height of 150 feet is the sweetgum (*Liquidambar styraciflua*). It is distinctive on two counts. Its star-shaped, five-pointed leaves are capable of a kaleidoscopic variety of autumn colors. These range from yellow, orange, crimson, red, and bronze, to purple. When the leaves drop, the ball-like fruits with projecting points remain suspended on slender stalks. These spiny balls hang from the twigs throughout the winter. The name of this tree relates to the sweet-smelling liquid which flows from the wood and congeals into a resin. Called liquid amber it was used in former years as a base for salves, soaps, perfumes, and chewing gum. In the South it was also used as a cure for Montezuma's revenge or dysentery.

Obviously all the unusual or even common trees of the Ozark-Ouachita region cannot be mentioned or described here. The area is a melting pot or center of convergence for trees of northern extraction, trees from the southern Appalachians, and trees of southern affinities. Many are important as sources of commercial timber and others are of value for erosion control and wildlife. For naturalists and outdoors people a ramble through these pleasant and varied forests is a rewarding experience.

The mountain forests of western North America are predominantly coniferous forests. These softwoods of the West are first encountered by travelers in the Black Hills of South Dakota and in the Rocky Mountains from Montana and Idaho to Colorado,

The big trees of the West, including sugar pines and Douglas firs, gradually shed their lower limbs and produce clear trunks that extend to eighty feet or more before the crown of needles is reached.

New Mexico and Arizona. In the Rockies the great zones of plant life succeed one another up the mountainsides until the tundra gives way to barren peaks, snow, and ice. Nowhere is there a better place to see the relationship of forests to environment than on these slopes that rise above the western edge of the midcontinental prairie and the eastern edge of the Great Basin desert.

The great bands of trees on the slopes of the Southern and Central Rockies begin in the foothills with oaks, junipers, and pinyon pines. Broadleaf trees, including willows, alders, birches, and cottonwoods are confined to the lower valleys and stream sides. At moderate elevations ponderosa pine becomes dominant. Its parklike groves, where the wind breathes with a resinous odor as it whispers through the forest, are favorite places for all people who enjoy the mountains. Near the upper edge of the ponderosa pine zone, Douglas fir mixes with the pines and becomes dominant on cool north slopes. From 7,000 to 9,000 feet the pines fade out and Douglas fir populates the mountainsides. Above 9,500 feet, spruce and alpine fir form dense cool groves. At elevations of 10,000 to 11,000 feet these trees dwindle in size and grow in grotesque stunted groups or mats that form the Krumholz, or "crooked wood" below the treeless alpine summits.

The bands or tree zones pictured in books are not always so apparent in nature. The montane zone of the lower elevations is characterized not only by mixtures of ponderosa pine and Douglas fir but is interspersed with open meadows, aspen groves, grassy hillsides, and rocky ridges. On rocky high points limber pine occurs and the colorful Colorado blue spruce grows near dashing streams in shady canyons. The Engelmann spruce-alpine fir zone, above the montane, likewise is interrupted by mountain meadows, lakes, marshes, and rocky outcrops where limber pine grows on areas exposed to the wind. Where forest fires have burned the spruce-fir forests, lodgepole pine grows in solid stands.

In the Northern Rocky Mountains, of Wyoming, Montana, and Idaho, ponderosa pine and Douglas fir grow abundantly at lower elevations. Above this zone, the Rocky Mountain fire type is prominent because of its extensive stands of lodgepole pine and aspen. Above this are forests of fir, hemlock, and larch. In northwestern Montana and northern Idaho, forests of western larch, western white pine, and fir cover the rugged terrain, the mixtures varying with altitude, exposure, soil, and precipitation. Because of the multitudinous environments created by these different forest types, a botanical crazyquilt of shrubs, grasses, and wildflowers adds to the attractiveness of the flora for anyone who embarks on a journey through these mountains.

Of all the pines you will see most frequently in the lower life zones of the western mountains, ponderosa or yellow pine (*Pinus ponderosa*) is the dominant one. Its inviting groves, savannalike at the edge of arid lands, cathedrallike on the floors of fertile valleys, and parklike on the rim of the Grand Canyon, are so clean of other trees and shrubs, so luminous and moderate in their shade, and so seductive with their resinous incense, that they are the favorite vacation lands of the West. From the Pine Ridge upland in the western Nebraska prairie to the basalt rocks and volcanic cinder cones of the Cascades of Oregon and Washington, you will find some of these pines holding aloft their dense crowns on mighty boles that sometimes reach 50 to 80 feet before the first wide-spreading branches appear. These giants now are few in number since the timber cutters have taken most of them for their valuable wood. You will know them by their bright yellow plated bark. Some of these plates are a foot wide and two or three feet long.

In their youth the saplings sometimes grow in almost impenetrable stands. Growth in the early years is slow, but competition for moisture by the extensive root systems ultimately kills most of the trees and allows the survivors to grow separately to form the parklike groves so characteristic of the ponderosa pine. For many years the bark of these young pines displays only dark ridges, a feature that gives them the name "blackjack." Not un-

Ponderosa pine seedlings invade grasslands. Eventually the shade of the trees and the mat of fallen needles reduces the grass and wildflower understory until the ground is almost barren of herbaceous plants.

til they reach an age of fifty years or more do they begin to produce the yellow plates that increase in size until the trees become patriarchs of the forest at ages of 150 years or more.

The leaves of the ponderosa pine come three in a bundle and grow from five to ten inches in length. They are dark green, flexible, and aromatic. A variety, *arizonica,* called the Arizona pine, produces needles usually five in a bundle, and cones that are stalked and non-prickly. The cones of the typical species are scarcely stalked and are armed with slender recurved prickles.

A very different kind of pine grows above the ponderosa pine zone from Alaska southward through the Rockies, on the east side of the Cascades, and in the Sierras in California. It is the lodgepole pine (*Pinus contorta* var. *latifolia*), noted for its gregariousness, its adaptation to fire, its prodigality of seedling production, and its exasperation for elk hunters. At high altitudes individual trees resemble other pines in outline; in groves at medium altitudes they grow as single species stands and in such competitive numbers that the trunks are nothing more than spindling telephone poles, fifty or sixty feet high and five or six inches in diameter. As lodge poles formerly used by the Indians the trunks of young trees were peeled of their bark when they were two inches in diameter and fifteen to twenty feet high.

The individuality of this tree arises from its habit of retaining its cones on the branches for many years. When fire destroys the stand the resinous scales open to release the unharmed seeds. The result is a new stand of thousands of seedlings which will grow into trees all of the same age. Competition is so fierce that many young trees die. But those that survive grow into a thicket of poles made more impenetrable by long-lasting branches and wind blown trees that fall into the tangle like jackstraws. Elk can run through these tangles with heads held high and antlers laid on their backs. Unlucky and unhappy is the hunter who kills and tries to extricate a 1,000 pound elk from one of these labyrinthine jungles.

The needles of lodgepole pine are dark green, about two inches long, and occur two in a bundle. They stand stiffly outward from the twigs and en masse give the appearance of prickly fox tails. The wood is soft, durable, and is used for fence posts, telephone poles, railroad ties, corral fences, mine props, fruit boxes and, even yet, wigwam poles.

Like the lodgepole pine, the quaking aspen (*Populus tremu-loides*) forests are merely phases in the successional process that leads to other longer lived species, including Engelmann spruce and alpine fir. After fires, when the ground has been swept clean, aspens, fireweeds, lupines, and kinnikinnick appear in profusion in the areas of greatest moisture. Aspen reproduction comes largely from root sprouts or sucker shoots.

Millions of people are familiar with this most widespread tree on the continent. Its smooth white or greenish trunks, its red resinous winter buds, its dangling catkins in early spring, and its shining green leaves on flattened petioles that allow them to tremble in the slightest breeze, make the aspen identifiable on sight. In autumn the aspen groves cover western mountainsides with luminescent gold and crimson in stark contrast with the somber green of pine, spruce, or fir forests.

In the Rockies at elevations up to 12,000 feet in the southern mountains, Engelmann spruce (*Picea engelmanni*) grows in dense forests of spirelike trees with drooping boughs that shed the winter snows at the first sign of melting. These forests are dark and cool even in summer. Their shade is so deep that few herbaceous or shrubby plants grow within the confines of the forest. Yet the seedlings of these spruces endure the shade of the parent trees and ultimately renew the stand, making it a truly climax forest.

An even more slender tree that grows to upper timberline is the subalpine fir (*Abies lasiocarpa*). Its deep purple cones in summer and its multiple tiers of stiff horizontal branches make it the darling of photographers. The needle sprays are so strong that snow does not spill from the flat whorls of branches. The snow has to melt to obtain release. These striking trees grow to heights of seventy feet in the Rockies, excepting at timberline where they grow to only a few feet. They reach their greatest magnificence in western Oregon and Washington and in British Columbia where they attain heights up to 175 feet.

The Pacific Mountain System is truly a land of tree variety. Its spectrum of tree combinations ranges from the elfin or chaparral forests of southern California and the Joshua tree groves on the border of the desert to the giant sequoias of the Sierras and the dripping rain forests of the Olympic Mountains. This land of extremes is enlivened by so many environments there is no reason to wonder why it supports so many trees from widely diverse plant families. In much of its area conifers are dominant. Junipers, pines, spruces, and firs cover thousands of square miles. But oaks and other broadleaf trees are also abundant. Some of these are localized in their distribution, others are unique in appearance, and many are specifically adapted to local soils and climates.

The Oregon white oak grows on the west slopes of the Cascades and north to British Columbia. This noble tree reaches a height of 80 feet and an age of 500 years. It is a valuable hardwood species for furniture, buildings, fuel, and shade. Its acorns furnish food for deer and other wildlife.

Additional variety results from the altitudinal differences between sea level, or the deserts, and the mountain tops some of which exceed 14,000 feet. Vegetation zones beginning in the southern desert area exhibit transitions from shadscale-sagebrush through juniper and pinyon pine to limber pine and bristlecone pine to treeless alpine summits. Farther north in the Oregon Cascades the transition is from sagebrush desert through junipers, ponderosa pine, to firs and spruces. Along the Pacific coast a veritable menagerie of unique and interesting trees occurs because of the environmental amplitude of north and south facing slopes and of differences near and remote from the ocean.

One of the oddities found on the Sierran slope and in the coastal forests is California torreya (*Torreya californica*), also called California nutmeg. This tree, an endemic, native only in California, is related to the conifers but produces hard berrylike fruits instead of typical cones. Its relative, Florida torreya or stinking cedar (*T. taxifolia*), is rare and local in southwestern Georgia and northwestern Florida. Years ago when I first visited the Apalachacola area I asked a venerable, bewhiskered gentleman how to find the stinking cedars. He replied, "You mean *Torreya taxifolia*, don't you?" Astounded, I asked how he happened to know the scientific name. "Years ago," he said, "Dr. Aza Grey of Harvard University came down here to study them. He made them famous and they still attract many botanical visitors."

I have seen the two Torreyas in America but there is one more I would like to see, *Torreya nucifera*. It is endemic on the Kamchatka Peninsula of northeastern Siberia between the Sea of Okhotsk and the Bering Sea. Fossils of ancient Torreyas have been found between this far off place and the present groves in California and Florida, indicating that the present endemics are remnants of a genus that once was widespread across the continent and into Asia.

Another tree, the only species of its genus in the West, is the California laurel (*Umbellularia californica*), found in the Coast Ranges of Oregon and California and in the southern Sierra Nevada. In Oregon it is called Oregon-myrtle, or just plain myrtlewood. Actually, the tree is neither a myrtle nor a true laurel, but a distinct genus in its own right. Its beautiful, easily worked wood polishes like marble and is made into bowls, trays, lamp stands, and other products by dozens of woodworking shops in southwestern Oregon. It is one of the highest priced of all hardwoods. History says that the golden spike at the celebration of the completion of the railroad at Promontory, Utah, was driven into a myrtlewood tie.

The golden chinkapin (*Castanopsis chrysophylla*) is a distinctive tree because of its spiny burs which contain edible nuts. The golden yellow scales on the under surfaces of the evergreen leathery leaves make it really different. And finally, creamy white flowers cover the pyramidal chinkapin trees like snow blankets in summer, making them conspicuous from afar on the forested hills. The trees grow from western Washington and Oregon south in the Coast Ranges to central California and locally in the Sierra Nevada.

There is no end to the distinctive features of other trees in the far western forests. The open cones of the incense-cedar (*Libocedrus decurrens*), for example, have been likened to Donald Duck's bill with his tongue sticking out. This is a lofty tree that grows from the southern Cascades in Oregon southward on the Coast Ranges of California and in the Sierra Nevada where it associates with the giant sequoias. It is the pencilwood tree from which pencils are made.

Some trees are notable because of bark characteristics. Tecate cypress (*Cupressus guadalupensis*), native in the Santa Ana Mountains, on Mount Tecate on the United States-Mexican boundary, and in Baja California, is significant as an ornamental since it can withstand heat, wind, poor soil, and the gardener's shears. But its most pleasing feature is the bark, covered with brownish scales which blow away in the wind and leave a smooth skin blotched with red and green with some resemblance to the bark of a sycamore. The Pacific madrone (*Arbutus menziesii*) is a tree made handsome by its orange red berries, its thick leathery evergreen leaves, and its reddish brown bark which is always peeling into papery tatters.

Possibly the most photogenic of all mountain trees is the bristlecone pine (*Pinus aristata*). It ranges from Colorado, New Mexico, and Nevada to California at elevations up to 12,300 feet. Subject to the dry winds of summer and the cold blasts of winter, growing on alkaline soils lacking in important minerals, surviving where snow may not melt even in July, the bristlecone pine with its exposed wood and grotesquely twisted trunk and branches achieves the longest life of any living tree. Even the giant sequoias cannot approach the bristlecone's life span of possibly more than 4,000 years.

The great age of the bristlecone pines has prompted an intensive study of their growth rings by means of increment borers which are long hollow drills used to remove a wood core without killing the tree. A tree ring is produced each year. This enables the investigator to count from the outside inward to the oldest ring in the center of the tree and thus determine its age. Since the rings vary in thickness according to the weather during the year of growth, weather patterns of the past are indicated.

When overlapping patterns of living trees are linked with those of older trees long dead, the chronology can be carried back thousands of years. Dates assigned by these tree rings have been used to correct discrepancies in the radio-carbon method of dating archaeological artifacts. The improved dating has resulted in revision of the dates of various human cultures in the distant past. For example, some of the tombs in Brittany now are dated as being 1,500 years older than the Egyptian pyramids.

The giant sequoias (*Sequoia gigantea*) of the Sierra Nevada and the redwoods (*S. sempervirens*) of the Pacific coast region are the true monarchs among trees of the world. The big trees of the Sierra are the largest plants on earth and the redwoods are among the tallest. Some of these giants exceed 2,000 years in age and heights of 370 feet. Many writers have grown lyrical in expressing their sense of awe and even reverence arising from the unbelievable solemnity and majesty of these forest giants. Only by visiting them and seeing their fluted columns towering above misty aisles to remote crowns in the sky can one appreciate their immensity. Here, if anywhere, time seems to stand still.

These, however, are not the only giants of the Pacific Mountain forests. Inland from the redwoods are the sugar pines (*Pinus lambertiana*) with their two-foot long cones. And from the Cascades west to the Olympic Mountains are forests of western hemlock (*Tsuga heterophylla*), grand fir (*Abies grandis*), western redcedar (*Thuja plicata*), Sitka spruce (*Picea sitchensis*), and enormous Douglas firs (*Pseudotsuga menziesii*). In these finest of forests the Douglas firs are giants in their own right. In the dripping jungles of the rain forest the profusion of trees, so huge they are almost beyond comprehension, astounds the viewer. Towering one hundred feet or more before the first branch is reached some ancient specimens reach fifteen feet or more in diameter. The tallest Douglas fir ever recorded was 385 feet high.

The call of "Timber!" in the Northwest, the whine of the

Sequoias, sugar pines, Coulter pines, and Douglas firs grow on millions of acres in the Sierra Nevada. Some of these trees reach heights of 200 feet or more if left to maturity.

Forest debris is not wasted in nature. Ants, termites, wood-boring beetles, and fungi gradually return it to earth, where it enriches the soil. These trees were blown down by high winds.

chain saw, and the insatiable demand for lumber, plywood, and other forest products has virtually eliminated the great virgin stands that once dimmed the forest floor. The cutover lands now produce second growth timber which also is being cut, even before it matures. If you want to see the Northwest wilderness as it once was you will have to go to Olympic National Park where the western redcedars, Douglas firs, Sitka spruces, and western hemlocks still are undisturbed by man.

Vanilla-leaf is a perennial herb that grows in the shade of forests in the
Northwest. The leaves and flower stalks arise from creeping rootstocks.
The palmately divided leaves have a vanillalike fragrance when dried.

6

Mountain Wildflowers

NOWHERE CAN YOU FIND a more colorful wildflower pageant than in the mountains. The flower season begins as early as February in the southern mountain ranges and moves northward into April or May as one approaches the Canadian border. Spring also moves upward on the slopes as the weather ameliorates, first in the foothills, later in the tree zones, and not until June or July when the snow melts in the alpine meadows and rock fields.

Summer is short at high altitudes and the wildflowers have only a few weeks to complete their growth cycles before autumn is in the air. Then the procession of flowering plants—goldenrods, asters, erigerons, and other yellow flowered composites—moves down the mountains to give the last colorful display in the foothills.

The autumn foliage display also is seasonally oriented by elevation. The dwarf birches and heaths evoke their best colors in the Presidential Range in New Hampshire in early fall. The oaks, maples, walnuts, sassafras, cherries, and hickories in the Appalachians are spectacular with their reds, browns, and yellows in October. Aspens splash the slopes with gold in the Rockies in early October and when the frosted leaves have fallen, the narrow leaf cottonwoods fringe the lower riversides with lemon yellow groves. In the Cascades and Coast Ranges of Oregon and Washington the vine maple glows in sunlight like red flames against the somber green background of Douglas firs. These are the spectacles of color en masse before winter dims the mountainsides to shades of forest gray and brown and green.

For the flowers of individual herbs, shrubs, and trees, each season has its merits. How we all thrill at seeing the first harbingers of spring. What pleasure we experience when seeing the jack-in-the-pulpit unfold its spathe, in finding the first bloodroot flowers beneath the still leafless shrubs, or picking a bouquet of Dutchman's breeches from the moist soil beneath the oak and basswood trees. Surely it is spring when the nodding flowers of yellow adder's tongue or of the white dog tooth violet—both of which are lilies—arise between their two smooth and shining flat leaves. And definitely it is spring when the flowering dogwoods adorn themselves with sprays of white petallike bracts surrounding the small greenish clustered flowers. They lend enchantment to the greening woods.

Summer has its attractions too. In the southern Appalachians the bush-honeysuckle (*Diervilla sessilifolia*) reveals its greenish-yellow flowers from June to August. The sourwood tree (*Oxydendrum arboreum*) blooms in the same months with showy cream colored flowers in one sided clusters. In western subalpine meadows the elephantella or little red elephant (*Pedicularis groenlandica*) covers acres of land with a sward of reddish-purple flowers, each simulating an elephant's head, trunk and all. These and a thousand others appear in the high mountains bearing exquisite brilliant blue, orange, purple, white, red, brown, or green blossoms. Summer is the time to visit the high altitudes to see this myriad of blossom colors for then the floral spectacle is at its best.

With the advent of fall a varied assortment of new flowers appears. Their colors are more flashy than those of the spring flowers with their delicate hues. The fall flowers display intense shades of red, orange, yellow, blue, and violet. The whites and pinks of spring are no longer dominant. In fall the wildflowers grow in meadows, along stream banks, and open woods where the sunlight has promoted the growth of leaves and stems in preparation for the flowering and fruiting season. This is the time of year when gentians, lobelias, asters, goldenrods, coneflowers, and rabbitbrushes put on their prolific dazzling pageant of color.

You may have noticed that the majority of spring wildflowers are low in stature and that many of them raise their blossoms on scapes or stalks before the leaves and stems are fully developed. Like the lilies, crocuses, and onions in our gardens, some of the early wildflowers grow from bulbs or corms in which the food for plant metabolism was stored in the previous growth season. They take care of their blooming period first, and then develop mature foliage for photosynthesis to capture the energy of the sun to store food for another year.

The summer and autumn wildflowers commonly make vigorous vegetative growth during the warm season. A great many of the perennials in the mountains actually begin active growth as soon as the snow melts. They continue to produce stems and leaves for several months. Some of these, such as the larkspurs, angelicas, cow parsnips, thistles, and sunflowers attain remarkable sizes—up to eight or ten feet in height—before they bloom. Because of their profusion of leaves and photosynthetic surface they can produce food in quantity for the developing fruits and seeds.

Wildflowers in the mountains provide an opportunity for fascinating study of periodicity of plant development stages in relation to elevation and climate. This study of time of appearance of buds, flowers, seeds, and other biological phenomena is called phenology. Many people find it interesting to keep yearly records of the time of flower appearance in their home localities so they can note whether it is an early, normal, or late spring or summer. Thomas Mikesell recorded phenological dates and weather between 1873 and 1912 at Wauseon, Ohio. This is somewhat of a record. But more interesting was the work of A. D. Hopkins who related periodic events to climate, temperature, and elevation.

Some years ago I participated in a study of trends in seasonal plant development as related to snowmelt at different altitudinal zones in Ephraim Canyon on the western slope of the Wasatch Plateau in Utah. The study area ranged from 6,500 feet in the oak zone to more than 10,000 feet in the subalpine spruce-fir zone. Grasses, forbs, and shrubs were observed for periodic phenomena such as date of snow disappearance, date flower buds were evident, date of full bloom, time of seed ripening, time of seed or fruit dissemination, and plant dried up or leaves all fallen.

The date at which a particular stage of growth was reached

Wild geraniums are common mountain plants. The perennial species have pink flowers and the clumps of stems arise from deep woody roots. These plants are palatable to livestock, deer, elk, and moose.

became progressively later with increase in altitude. Based on averages for ten years, sticky geranium (*Geranium viscosissimum*), for example, had flowers in bloom on June 14 at 7,150 feet; on June 26 at 8,450 feet; and on July 20 at 10,100 feet. For the same elevations, average blooming dates for tongueleaf violet (*Viola linguaefolia*) were May 14, May 29, and June 23. Letterman needlegrass, western yarrow, common serviceberry, mountain snowberry, and other species each had their own rates of delay. But, in general, the appearance of development stages such as leaf buds bursting or flowers in bloom were about 12 days later for each 1,000 feet increase in elevation. Thus to see geraniums in bloom for five or six weeks continuously one would have to go to 1,000 feet higher every 12 to 14 days.

This study revealed many other aspects of plant growth on mountainsides. Each grass, forb, and shrub species had its own rate of delay with increase in elevation. Delay in dates of development were not always proportional to increase in elevation. These variations were due in part to differences in slope and exposure. Total height growth was greater at low than at high elevations, partly due to the short growing season in the subalpine zone. Growth, however, was more rapid at higher than at lower elevations, once it started. This is the result of growth beginning earlier in the season at low altitudes, when the plants are subjected to unsettled weather conditions and low average temperatures. At higher elevations growth inception is delayed until the snow cover is gone; by that time, higher seasonal temperatures prevail and the plants compress their stages of development into a shorter growing period.

The date when winter snow disappears from mountain slopes and valleys fluctuates widely from year to year. The disappearance rate as the snow line recedes up the mountain is determined principally by the depth of snow accumulation during the winter and by wind movement and temperatures during the period of melting. Since active growth of most wildflowers does not begin until snow has disappeared, the flowering season is correspondingly early, late, or normal. Variation in depth of accumulation at different altitudes and on slopes facing in different directions may result in early seasons in one place and late seasons in another place.

If one pays attention to environmental influences and is aware of the inherent periodicity of plant development, then the observation of wildflowers in the mountains becomes more than an excursion just to find and name plants. It leads to better understanding of why plants are where they are, and how they live their lives.

In the eastern mountains, from the Katahdin region in Maine to the southern end of the Appalachians, the flowering season begins with a few choice species that are harbingers of spring. Many of the interesting plants appear early on the forest floor where they complete much of their life cycles before the tree leaves shade them from the northward march of the sun. By mid-June the heaths reach their climax of spectacular flower abundance and variety of color. After this the summer and early autumn blooming plants provide an ever changing display of species until the killing frosts begin as early as late August in the north and advance southward until winter begins in the southern mountains in late October or in November.

The list of interesting plants in this great band of mountains is almost endless. If you count the trees, many of which have flowers of surpassing beauty, the flowering shrubs, the herbs or forbs, and the plants which produce no colorful flowers—sedges, grasses, ferns, lichens, mosses, and even mushrooms—you will find their numbers running into the thousands. In a single region such as the Great Smoky Mountains more than a thousand kinds of flowering plants occur. The number decreases as you go northward to mountains with more rigorous climates. Fewer species also live on mountain tops where the short growing season excludes many of the less hardy plants of the middle and lower mountain slopes.

Learning the more common plants and associating them with their habitats is a worthwhile and pleasurable pursuit. Some outdoor enthusiasts specialize in knowing the plants in specific localities such as the alpine zone or in mountain bogs where many species are unique and are found in no other habitats. Other people specialize in certain plant groups such as the orchids, the ferns, or the rhododendrons and azaleas. The dedicated pursuit of any one of these groups of plants can lead you into strange places in search of specimens you have not seen or added to your list. It can also acquaint you with the kaleidoscopic variety of plants as they are influenced by soil, altitude, the seasons, and the inherent growth characteristics of the plants themselves.

Each mountain zone has its characteristic plants which you learn to greet as old friends after frequent visits. On Mount Marcy's summit in the Adirondacks the bluets, violets, and dwarf buttercups grow in colorful carpets. The white bell-shaped flowers of diapensia (*Diapensia lapponica*), conspicuous in the ground-hugging mats of bright evergreen leaves, are there to greet the early season mountain climber. In June the hand-

Western skunk cabbage displays its golden yellow hood-shaped spathe enclosing the fleshy spadix in moist boggy areas below 3,000 feet in the conifer forests near the Oregon coast. It also is common around the base of Mount Rainier in Washington.

some roselike flowers of Lapland rosebay (*Rhododendron lapponicum*) is one of the showy species. Black crowberry (*Empetrum nigrum*) is another dwarf evergreen, common on northern peaks and also on the rocky coast of the Atlantic Ocean. The flowers, which have no petals, are small and inconspicuous. But the black berrylike fruits are prominent in autumn. Another creeping dwarf is the bearberry willow (*Salix uva-ursi*), found among rocks and in crevices in the alpine tundra. Its catkins are pink and the capsules are smooth brown.

The heath family includes a number of interesting shrubs, many with evergreen leaves, adapted to high altitudes, extreme weather conditions, and acid soils. Leather-leaf (*Chamaedaphne calyculata*), an early bloomer, is typical in bogs and areas of wetness and seepage. The white urn-shaped flowers are pendant, like strings of little bells on the slender twigs. The flat rusty green leaves remain over winter and are ready to perform their photosynthetic function at the first favorable temperature. Other common heaths of boggy areas are the pale laurel (*Kalmia polifolia*) and sheep laurel (*K. augustifolia*). The latter is poisonous to livestock and hence is also known as lambkill. Its flowers are pink to red and are borne on the twig below a terminal cluster of leaves. Pale laurel produces its pink cuplike flowers in terminal clusters on the twigs. Another species, mountain laurel (*K. latifolia*) grows abundantly in the Berkshire and Catskill Mountains and southward in the Appalachians. Labrador tea (*Ledum groenlandicum*) is another distinctive heath, common in bogs and peaty areas. The leaf margins are rolled at the edges and the under surfaces are covered with woolly felt, white at first and brown as the leaves mature.

The Vacciniums are an interesting group, especially those with palatable fruits. The low sweet blueberry (*Vaccinium angustifolium*) with the white or pink tinged globular bells grows from the seashore to the mountain tops. The dwarf variety *laevifolium* grows on alpine summits while the taller variety *angustifolium* is characteristic of the lowlands. The berries are blue. Its relative the bog bilberry (*V. uliginosum*) has oval blue-green leaves, red flowers, and blue to black berries.

A plant I always examine with pleasure is the twinflower (*Linnaea borealis*). It must have been a favorite of Carl von Linnaeus, the Swedish botanist who promulgated the binominal system of naming plants. The twinned flowers are appropriate to the system of giving two names to plants and animals. The white bell-shaped flowers that rise from vinelike stems droop from a single stalk .that branches into two slender pedicels. The five-lobed corolla, tinged with pink and marked with rose-purple stripes, produces a delicate fragrance. Twinflowers are common in cold bogs from Labrador to New Jersey and in the mountains south to Pennsylvania and Maryland. I have found them in moist woods in Michigan and in the mountains in Oregon and Washington.

There are three small plants with conspicuous white flowers that are worth a second look. Three-toothed cinquefoil (*Potentilla tridentata*) has creeping stems and white flowers; its relatives have yellow flowers, including the upright shrubby cinquefoil (*P. fruticosa*) which grows from Labrador to Alaska, south to Tennessee, in my front yard in Fort Collins, Colorado, and throughout the Rocky Mountains. Mountain sandwort (*Arenaria groenlandica*) is a high altitude tufted plant with conspicuous white flowers. It has several relatives in the mountains of the

West. Dwarf cornel (*Cornus canadensis*) is a pleasing little ground loving dogwood, common on Adirondack peaks and westward in damp woods to Minnesota and Alaska. Like the flowering dogwood its cluster of tiny greenish-white flowers are surrounded by four broad white bracts that many people assume are petals. In early autumn the cluster of bright red berries explains its other name, bunchberry. Its small stature, four to eight inches high, puts it into contrast with its shrubby relatives such as the red-osier dogwood and the tree-sized flowering dogwood.

If you walk far along the Appalachian Trail you will see a profusion of other flowers, including the common species such as trilliums, yellow and purple violets, wild onions, serviceberries, May apples, and ferns with their fiddlenecks opening in early spring. An abundant herb that always brings admiration is the pearly everlasting (*Anaphalis margaritacea*). The plants are woolly and the parchment-like scales or involucral bracts give the appearance of shining white flower heads. The western variety, *occidentalis*, used to fringe the pavement on the pass over Mt. Hood, in Oregon, until the Highway Department sprayed the plants with poison. All they accomplished was erosion of the road shoulder and removal of several leagues of roadside beauty.

One other plant you should look for in the mountains of New England and New York is the alpine goldenrod (*Solidago cutleri*). It cannot compare in size with the giant goldenrods of the prairies in Wisconsin or Nebraska; instead it is one of the smallest goldenrods, usually less than ten inches high and possessed of a single upright stem. The bright yellow flowers are distinctive among the plants that grow on high Northeastern mountain peaks.

The outstanding and astonishingly beautiful flowers of the Appalachians are the rhododendrons. These shrubs produce some of the most colorful natural gardens in the world when they contribute their purple, white, organge, and magenta to the mountainsides. Some are large and some are small. The alpine azalea (*Loiseleuria procumbens*), as the Latin specific name indicates, is a depressed shrubby evergreen. It has tiny bell-shaped white or rose colored flowers in clusters of two to five. Look for it in alpine areas in Maine and New Hampshire.

The tall rhododendrons make their great splashes of color on the mountainsides in the southern Appalachians. Catawba rhododendron (*Rhododendron catawbiense*) with its lilac-purple broadly bell-shaped flowers is one of the conspicuous shrubs of the high mountains from Virginia to Georgia. Even more brilliant is the flame azalea (*R. calendulaceum*) which reaches small tree size in favorable sites. Its large orange blossoms burst as the leaves appear and then turn to flame color. Hybrids show many color variations including white, yellow, salmon, pink, orange, red, and deep red. In the Great Smokies mountain laurel (*Kalmia latifolia*) is an attractive evergreen shrub that superficially resembles some of the rhododendrons. Its pink or white flowers, borne in terminal corymbs, are broadly bell-shaped. In full bloom the shrub is a mass of color in the hardwood forests.

If you are in the southern Appalachians in late spring, do not neglect the tree flowers. Even the maple flowers are exquisite when examined minutely. A tree with distinctive flowers is the sourwood (*Oxydendrum arboreum*). It is also called lily-of-the-valley tree because of its compound clusters of creamy white blossoms resembling lilies of the valley. The southern Appalachian farmers know that bees make the very finest honey from these flowers.

Another favorite tree is the silverbell (*Halesia carolina*). It grows in the moist woodlands of the southern Appalachians and westward to Arkansas. Its variety, *monticola*, the mountain silverbell occurs in the southern portion of the Blue Ridge Parkway and is abundant in the Great Smoky Mountains. The white bell-shaped flowers, sometimes tinged with pink, appear in such profusion they splash the mountainsides with zones of white.

Although rare the yellowwood (*Cladrastis lutea*) is another tree worth seeing. Its bark is smooth silvery gray or blackish. The bright yellow wood formerly was used for dye by the hillfolk who spun their own cloth. But the most attractive feature is the drooping inflorescence of white pea-like flowers with prominent clawed and winged petals. This handsome tree grows locally from western North Carolina to southern Illinois and south to northern Alabama and extreme northern Georgia. You may also see it in Missouri, Arkansas and northeastern Oklahoma.

In the western mountains—the Rockies, Sierra Nevada, Cascades, or the Coast Ranges—you can stand almost anywhere in the growing season and see a pleasing variety of plants with showy flowers. Some of these are trees and shrubs as well as herbaceous species. The striking beauty of many of these mountain wildflowers impels even the average visitor to learn the names of at least some of these.

If you are not a botanist, skilled in the identification of plants by use of keys in plant manuals, you still can learn the names of the most interesting species with the aid of picture books and paperback guides sold at bookstores and in the western National Parks. *A Guide to Rocky Mountain Wildflowers* by John J. Craighead, Frank C. Craighead, Jr., and Ray J. Davis describes more than 590 species likely to be encountered in the mountains from northern Arizona and New Mexico to British Columbia. This is one of the Peterson Field Guide Series. An added feature is the inclusion of "interesting facts" for each plant species which correlate flowering dates with animal activities, environmental information, and altitude effects.

Other manuals and illustrated flower guides are available for almost any mountain area in the country to help you name the plants. Booklets for flower identification, especially relating to the flora of National Parks, are available for a nominal price. Some of these group the flowers by color as an aid to those not trained in botany. Identification of conspicuous plants thus is made easy and is the first step toward enjoyment and appreciation of the mountain flora. Complete satisfaction in nature observation, however, comes from further study and understanding of how wildflowers fit into their environments.

If you travel much through the mountains of the West you will become aware of the relationships between altitude and plant zones. Certain kinds of trees, shrubs, and wildflowers belong at certain altitudes. When you learn to recognize some of the common and abundant plants you will be able to make a good guess as to how high you are on the mountainside. In a general way, increase in altitude also corresponds to increase in latitude. If, for example, you climb from Colorado Springs to the summit of Pikes Peak you will be traveling the equivalent of 2,500 to 3,000 miles north, insofar as plant zones are concerned.

Vegetation in the life zones follows a general pattern in the western mountains but there are local differences, depending on where you are—Colorado, Utah, Arizona, California, Washington, or Montana—and on which side of the mountains you are visiting. If you approach the Rockies from the grasslands of the Great Plains you will meet the foothills zone at 6,000 to 8,000 feet. Here, shrubs such as mountain mahogany, three-leaf sumac, and scrub oaks occur in extensive stands. Plants from the plains and from the mountains mingle here in colorful shows of penstemons, gaillardias, wallflowers, pasque flowers, chokecherries,

Gaillardias, or blanket flowers, grow in dry soil from the mountain foothills to 8,000 feet from Canada south to Arizona and New Mexico. These colorful flowers have been domesticated and grow nicely in home gardens.

loco weeds, pussytoes, showy milkweeds, and perennial sunflowers.

The mountain forest zones begin above the foothills with parklike groves of ponderosa pine. The pines mingle with Douglas fir on north slopes. At still higher elevations Engelmann spruce covers north slopes in almost pure stands; elsewhere they mingle with alpine fir. In the subalpine zone, limber and bristlecone pines grow on the wind blown slopes below the treeless tundra. In this mountain conifer zone are stands of lodgepole pine, groves of aspen, and open meadow lands. Each of these provides a different environment where adapted wildflowers bloom.

In these forests some of the wildflowers reach perfection. Columbines are the notable ones with their lavender and white flowers. Yarrow seems to thrive best here along with Indian paintbrushes, meadow rue, asters, penstemons, and tall larkspurs. In the meadows there are fields of asters, louseworts, potentillas, gentians, shooting stars, and a multitude of other flowers.

On the approach to the Rockies or the great Colorado Plateau from the west you will leave the world of sagebrush, greasewood, and saltbushes of the semi-desert and enter the pinyon-juniper zone. In Utah, the transition from desert to forest is characterized by oak groves or by stands of pines and junipers. Pinyons (*Pinus edulis*) are intermingled with Rocky Mountain junipers (*Juniperus scopulorum*), one-seed juniper (*J. monosperma*), or Utah juniper (*J. osteosperma*). Wildflowers are not abundant in this zone, owing to dry site conditions.

Species with showy flowers include: gumweed (*Grindelia squarrosa*), a yellow flowered composite with sticky heads; scarlet falsemallow (*Sphaeralcea coccinea*), with salmon-red hollyhock-like flowers; various daisies (*Erigeron spp.*); and narrowleaf sunflower (*Helianthus petiolaris*), a small sunflower also common in the short-grass prairie. The upper edge of the pinyon-pine zone on both sides of the Uncompahgre Plateau in southwestern Colorado contacts the mountain brush community in which Gambel oak (*Quercus gambelii*), saskatoon serviceberry (*Amelanchief alnifolia*), and mountain mahogany (*Cercocarpus montanus*) are prominent. The latter produces spiral hairy seed plumes several inches long which give the shrub a feathery aspect in early autumn.

The life zones on the west side of the Sierra Nevada differ

Lodgepole pine needles are dark green, stiff, and two to a bundle. The orange red male flowers are crowded in short spikes. These trees may take a century to reach a height of sixty feet.

from those of the Rockies because of the gradual slope above the Great Central Valley in California. When white men first arrived the valley was covered with a magnificent prairie in which purple needlegrass (*Stipa pulchra*) was the principal grass. Now in the foothills bordering the valley are blue oaks, digger pines, and live oaks. Above these are forests of ponderosa pine, sugar pine, giant sequoia and lodgepole pine. Wildflowers to look for include: Sierra mule ears (*Wyethia mollis*), a composite with sunflowerlike heads and woolly-white hair on the leaves; showy penstemon (*Penstemon speciosus*), with bright purple flower clusters; little elephant head (*Pedicularis attollens*) which is similar to the elephant head (*P. groenlandica*) so common in the Rockies; and thimbleberry (*Rubus parviflorus*), a shrub with large white flowers and sweet red berries. Many other shrubs with colorful flowers also occur in this zone, including the manzanitas, gooseberries, sagebrushes, and the mountain whitethorn (*Ceanothus cordulatus*).

The subalpine forest occurs above 8,500 feet and extends to timberline. Here the important trees are mountain hemlock, western white pine, and whitebark pine. Many of the wildflowers in this highest forest also are seen in the high alpine meadows. Among the interesting ones are Sierra penstemon (*Penstemon heterodoxus*), purple mountain heather (*Phyllodoce Breweri*), Douglas phlox (*Phlox douglassii*), and the Sierra stonecrop (*Sedum obtusatum*).

Even if you do not know the names of individual wildflowers you can recognize many of the families or other groups to which they belong. The penstemons, for example, grow over a wide range of altitudes and generally have showy flowers in racemes, panicles, or cymes. The tubular corollas are often inflated and are two-lipped somewhat in the manner of the familiar snapdragons in our home gardens. The species are numerous; the Forest Service checklist for the Intermountain region of Utah, Wyoming, Idaho, and Nevada lists 99 different kinds.

Some of the penstemon species are widely variable and present a challenge to the specialist. Widely distributed ones such as Rydberg penstemon (*Penstemon rydbergia*) are easy to recognize. The broad range in flower colors makes this group an interesting one to study. Some of their names indicate their individual features—stickystem penstemon (*P. glandulosus*), yellow penstemon (*P. confertus*), beardlip penstemon (*P. barbatus*), Wasatch penstemon (*P. cyanathus*), and littleflower penstemon (*P. procerus*). The penstemons exhibit various growth habits, ranging from mat forms, to erect herbaceous plants, to much branched woody stems. *Penstemon Cardwellii* in the Cascade Mountains and at high elevations in the Coast Range of Oregon is a shrub with thick leathery leaves and deep purple flowers one to one and a half inches long. The calyx is hairy and the stamens are woolly. In the Front Range of the Rockies dark penstemon (*P. Whippleanus*) is distinguished by the very dark reddish purple flowers. It grows in the subalpine and timberline zone.

The Indian paintbrushes seem to attract everyone who sees them. These beautiful plants, abundant in the western mountains, owe their beauty to colored bracts, or modified leaves, instead of the flowers which are inconspicuous. In the Front Range of the Rockies you are sure to see a number of these showy plants. In the subalpine zones look for western paintedcup (*Castilleja occidentalis*) with its cream to sulfur yellow bracts.

The splitleaf painted-cup (*C. rhexifolia*) has bracts that vary from yellow to rose to bright purple. Another common species is the Wyoming paintbrush (*C. linariafolia*), with narrow leaves and flowers with green corollas that extend beyond the red sepals and bracts. A dwarf species, short-flowered paintbrush (*C. puberula*), grows in the alpine zone. The bracts are brownish or yellowish and are covered with long hairs.

The paintbrushes have many common names. Yellow paintbrush (*C. sulphurea*) is also called Indian paintbrush, painted cup, squaw-feather, and sulphur painted-cup. You can take your choice. It usually is best to stick with the scientific names, but in this case even the botanists disagree because the species are difficult to differentiate. The nearly 200 species that grow in the West, however, are easy to recognize as paintbrushes. Their colorful spikes make them among the most attractive flowers in the mountains.

Another group of plants, plagued by a confusion of common names, is the huckleberry or blueberry clan. The *Vaccinium* species in the Pacific Northwest are called huckleberries but in New England they are called blueberries. Also, in the East, species of the genus *Gaylussacia* are called huckleberries. Harlan P. Kelsey and William A. Dayton in the 1942 issue of *Standardized Plant Names* did not help the matter by listing the western species as "blueberries," "bilberries," and "whortleberries."

By whatever name you call them, including "grouseberries," they are of botanical, ecological, and commercial interest. The fruits of some are delicious and are grown for the market, especially in the East. Berry pickers spend hundreds of thousands of hours harvesting them in the mountains. Of added interest is the possibility that a bear will be picking berries in the same patch.

In the Rocky Mountains, fire in the spruce-fir forests generally starts a plant succession that progresses from a moss, grass, forb cover through lodgepole pine or aspen and ultimately back to spruce-fir. If the fire is light the ground soon may be covered with a thick mat of grouseberry (*Vaccinium scoparium*), dwarf bilberry (*V. myrtillus*), or dwarf huckleberry (*V. caespitosum*). The dwarf huckleberries do not seriously impede regeneration of the forest although they may persist as a bright green carpet beneath the lodgepole pines for many years.

The flowers of the huckleberries are small, urn-shaped to bell-shaped, and are inconspicuous among the evergreen leaves. The

The woolly leaves and tall spikes of yellow flowers of mullein are familiar sights along road shoulders in the mountains, in cutover lands, and open forests. Mourning doves frequent mullein patches in autumn and small birds cling to the dried stems in search of seeds above the snow in winter.

berries of red huckleberry (*V. parviflorum*), grouseberry, and dwarf bilberry are red. Other species in the Pacific Northwest are blue or black. The most frequently harvested species is the thin-leaved huckleberry (*V. membranaceum*), a two to five foot high shrub with blue-black fruits which may be nearly one-half inch in diameter. The small cranberry (*V. Oxycoccos*), which is picked from Maine to North Carolina, is a creeping vine like the cultivated cranberry. The mountain cranberry (*V. Vitis-Idaea* var. *minus*) also produces fruits which make delectable jelly.

Owing to their wide distribution and abundance in some mountain areas the Vacciniums make excellent ground cover and protect the soil from erosion. They also are valuable in the food chain. Many birds, including grouse, ptarmigan, and magpies eat the berries. Mice, rabbits, and coyotes relish them. And bears consume great quantities of both leaves and fruits. Deer eat the leaves, especially when snow covers low-growing shrubs and forbs.

A distinguished relative of the Vacciniums is the box huckleberry (*Gaylussacia brachycera*) of the wooded hills in Virginia, Pennsylvania, and West Virginia. The evergreen leaves resemble those of the box (*Buxus*), an ornamental, and the fruits are sweet berrylike drupes with 10 seedlike nutlets. The plants spread mainly through root growth; through centuries of time a single plant, if undisturbed by fire or other disasters, may cover an acre or more. Dr. Henry C. Cowles, in his lectures to ecology students at the University of Chicago, used to tell us that the box huckleberry could be the oldest living thing on earth. Based on the slow rate of spread a single large clump could be 6,000 years old. This could exceed even the life span of the California bristlecone pine.

Even the perennial wildflowers may live for a great many years if they do not suffer a pathic event. One of my favorites, which I have watched for many years, is a clump of coneflowers (*Ratibida columnifera*) in the short-grass prairie east of the Rocky Mountains in Colorado. It survived the drought of 1939, severe overgrazing by cattle in 1952, and trampling by well diggers in 1966. In the early summer of 1973 it was still there with five flower stalks.

In 1941, I marked a square meter with iron stakes around three clumps of blue grama (*Bouteloua gracilis*) in the ponderosa pine forest on the Rampart Range north of Woodland Park, Colorado. That was a year of good rainfall and the forest floor was colorful with goldenpea or golden banner (*Thermopsis divaricarpa*), a plant with pealike blossoms and three-leaved foliage. In 1970, when I visited the plot again, the grass clumps were there, slightly changed in shape, but demonstrating how perennial grasses may live for many years. The golden peas were not so numerous but I do not doubt that many of them still grew from persistent creeping rootstocks.

Botanists sometimes speak of the area west of the Rockies as "lupine country." These pleasing plants are common in moist mountain forests, open pine woods, and on poor shallow soil. In the latter habitat they are soil builders through the nodules of their roots which are able to take nitrogen from the air. Near at hand most of the lupines are handsome plants. In favorable years they also decorate the landscape with spectacular clumps or masses of blue, white, or yellow flowers.

The genus is easy to recognize. The palmate leaves open with numerous leaflets spread into a symmetrical circle. The papilionaceous or pealike flowers are precisely arranged on upright spikes and the colors are vivid. Dozens of species are found throughout the West, but unfortunately they are difficult to distinguish. Unless you are a trained botanist it is best to content yourself with knowing you are seeing lupines.

One common species in the Blue Mountains in eastern Oregon, sulfur lupine (*Lupinus sulphureus*), has bright sulfur-yellow flowers on densely tufted stems. On cut-over land or in dry open ponderosa pine forests it is a photographer's dream when one of these clumps with spikes of gold appears against a fallen log or a tree stump. Few of the lupines have white flowers but the whitish lupine (*L. sericeus sericeus*) does. It also grows in the Blue Mountains in Oregon.

A fairly common lupine in the Rocky Mountains is the lodgepole lupine (*L. parviflorus*). Its blue flowers may be seen from late June to early August in most of the elevational zones. The low lupine (*L. pusillus*) is an annual plant, interesting because of its low stature. The plants seldom grow more than two to eight inches high. The leaves usually are smooth above but may be long-hairy beneath. The flower racemes are only one to two inches long but the purplish, bluish, or sometimes white flowers make it a pleasing plant to see in the foothills and on sandy slopes up to 8,000 feet. It grows from Washington to Arizona and east to Kansas.

There are many reasons why people enjoy wildflowers. One reason is that mountain wildflowers grow in enchanting places. These environments vary from heath balds, which are areas barren of trees, on high Appalachian peaks to shaded canyons where ferns make exuberant growth. The desert mountains cast a different spell as one wanders among the saguaro cacti or century plants with their bizarre and interesting shapes. Then there are the semi-aquatic plants, such as the monkeyflowers, which grow in and beside rushing mountain streams.

Association of wildflowers, such as these water-loving plants, with other activities sometimes doubles the pleasure of doing things in the mountains. For example, I tend to associate monkeyflowers with fish. When these colorful members of the figwort family are in bloom the cutthroat trout, the rainbows, and even the elusive Dolly Varden trout are active. Some of my best catches of big trout have been among the monkeyflowers in cold mountain streams. So, when I think of trout I think of monkeyflowers, and sometimes vice versa.

There are several species of these plants but two are common and widely distributed. The yellow monkeyflower (*Mimulus guttatus*), with bright yellow flowers resembling snapdragons has red spots in the throat of the corolla. The red monkeyflower (*M. lewisii*) has yellow spots in the corolla throat. The stigma lobes are equipped to close and hold pollen as a bee backs out of the flower, thus preventing self-pollination. The square stems are hollow as befits a plant that grows in wet places and needs aeration of the internal tissues of the plant. Monkeyflowers flourish at the edges of springs, in seepage areas, and even in the cold gin-clear water of rushing mountain streams.

Among the floral curiosities in the mountain woods are the showy saprophytes which appear in white, or in pleasing tints of scarlet, red, purplish, or chestnut brown. The Indian pipe (*Monotropa uniflora*) is the ghostly one with clustered, waxy stems, four to ten inches high, and leaves that are merely white scales. The drooping pipelike form with the nodding flowers is common in the eastern United States as well as in the mountains in the West. It has no chlorophyll and obtains its nourishment from decayed organic matter in the soil.

Sugar stick (*Allotropa virgata*) is another showy saprophyte in far western forests. The pure white stem is striped with scarlet and resembles a two-toned barber pole when the scales are removed. The flowers have no petals but the sepals are brightly colored. The urn-shaped flowers stand erect in a raceme. The plant is not as tall as its relative, pine drops (*Pterospora andromedea*), which is a simple wandlike red stem that sometimes

grows to a height of three feet. The inflorescence is a long raceme with many nodding white or reddish flowers.

Another curious form found in the Pacific Coast forests is
the cone plant (*Newberrya congesta*). It can be mistaken for a
cauliflowerlike fungus when it pushes through the forest floor to
form a mound of white or brownish stems several inches in
diameter. The flowers in the crown of the plant are white or
flesh colored and waxy. Still another of these saprophytes is
hairy pine-sap (*Hypopitys lanuginosa*) with a reddish or yellowish stem that reaches a foot or more in height.

The parasitic plants also are widely distributed and frequently
attract notice even if they do not have showy flowers. The mis-
tletoes parasitize various trees including pines, spruces, larches,
and oaks. The basket sized clusters of fragile yellowish stems in
the crowns of trees are unmistakable. The small, usually greenish-yellow flowers are clustered in spikes or panicles. The small
bluish or purplish fruits burst at maturity and expel the seeds
which stick to other tree branches or are carried on bird feet to
start new plants on other trees.

Some of the parasitic plants derive their nourishment from the
roots of herbs and shrubs. Some years ago I received a package
without name or address. Inside was a brownish plant, mostly
rootstock, with yellowish two-lipped flowers on a short stout

The fruiting bodies of fungi appear on fallen trees after the mycelium
has penetrated the wood and accumulated food material. These form
the mushrooms with their spores which disseminate the plants.

stem. Also enclosed was a note which read, "Dave, what in hell is this?" I rightly guessed that it came from my old friend, Ed Miller, forest supervisor at Cody, Wyoming. My reply was equally short, "Broom-rape (*Thalesia fasciculata*), parasitic on sagebrush." Later the botanists changed the name to *Orbanche fasciculata.*

Some people grow amaryllis flowers at home, but a really big one is the century plant (*Agave scabra*) that grows in the Chisos Mountains in Big Bend National Park in Texas. Like most century plants it spends ten to twenty years storing food in its thick, fleshy, spiny toothed leaves. Then in a burst of growth it quickly sends up a mighty flowering stalk fifteen feet high with showy flowers that stand out from the stemlike epaulets on leafless branches. When the seeds are formed the mother plant dies. But offsets around the base continue to grow and perpetuate the species.

If you want to go from the gargantuan to the minute, look for the tiny butterwort (*Pinguicula vulgaris*) when you are clambering over wet calcareous rocks or trotting around bogs. It is possible to trample hundreds of them into the mud without knowing they are there unless you observe closely. They belong to the bladderwort family and are small, stemless perennials with greasy leaves that catch small insects in their mucilaginous covering. The solitary flower is borne on a one or two inch high leafless stalk. The lips of the corolla are usually violet in color. These plants grow in New England, west to Wisconsin and Minnesota, and in moist places in Oregon and Washington.

Bogs and muskegs are among the most interesting of mountain habitats. They occur all across the northern reaches of the continent, down the Appalachians, and southward from Canada to the Cascades of Washington and Oregon. One of my favorites is Gold Lake Bog a few miles south of Waldo Lake in the Cascades of southern Oregon. Part of the area is floating mat and part is open water. When I last visited the bog and its vicinity it was one of the finest examples of unspoiled mountain lake, bog, meadow, and surrounding subalpine forests. Aquatic and emergent plants were involved in the usual successional series from aquatic to terrestrial systems.

Among the sedges and grasses was an old friend I had known in the bogs of Wisconsin and Michigan, the buck-bean (*Meny-*

anthes trifoliata). The white and rose flowers were grouped, a dozen or more, on each raceme and the trifoliate leaves with long petioles rising from thick creeping rootstocks made the plants unmistakable, even though I had not seen the species for more than twenty years. Cotton grass (*Eriophorum gracile*) and pale laurel or bog laurel (*Kalmia polifolia*) also were there.

Of greatest interest was the presence of all five carnivorous plant species that grow in the Oregon Cascades. The three bladderworts, common bladderwort (*Utricularia vulgaris*), mountain bladderwort (*U. intermedia*), and smaller bladderwort (*U. minor*) were growing together. The little bladders which float these plants during the flowering period are furnished with tiny valvular lids that open so minute animals may enter. Bristles around the pores, when touched by small swimming organisms, actuate the sacs or bladders to expand and suck in the tiny creatures. These are digested and become a part of the plant's nourishment.

The round-leaved sundew (*Drosera rotundifolia*) and the long leaved sundew (*D. longifolia*) also were present. These tiny plants with rosettes of brownish leaves are equipped to act as insect traps. Sensitive hairs on the leaves are tipped with a glandular secretion that holds the insects fast while the blade gradually folds and digests its catch. When the job is done the leaf unfolds and the trap is set for another victim.

In the open water of Gold Lake I saw clouds of tadpoles rising to the surface and submerging at the edge of the sedge mat. These came from the eggs of two species of frogs, *Rana cascadae* and *Rana pretiosa* which live there in close association. And of all the most unlikely animals I ever expect to see in a bog, there was a badger, undoubtedly searching for the frogs. It demonstrated what a fascinating place for study of geologic, pedologic, and biologic dynamics a mountain bog can be.

Baby porcupines and adults eat flowers and other herbaceous plant materials in summer. When winter comes they subsist on bark of trees, including either conifers or hardwoods. If snow is deep they may remain in one tree for weeks at a time.

7

Mammals of the Mountains

THE MAMMALS OF THE MOUNTAINS are a varied lot since the diversified landscapes and life zones provide a myriad of habitats. The many forest types, the rocky outcrops, the northern and southern slope exposures, the streams, ponds, lakes, and meadows, and the latitudinal extent of the great mountain chains provide environments for more varied populations than ever could exist on the prairies or in the deserts.

Some of the mammals are so common they may be seen almost daily. Practically anywhere one species or another of the chipmunks will be actively engaged in food gathering during the summer months. Deer are common and frequently emerge from forest cover at dusk to graze in mountain meadows. In some localities bears are common. They are symbols of great populations that once inhabited the prairies and the lowlands but have been driven to the fastness of the high forests by encroachment of modern civilization.

The mammals we seldom see are the very numerous small creatures—mice, shrews, moles, and gophers—that work under cover of dense vegetation, burrow into the earth, or live nocturnal lives. The daytime mammals are more frequently seen, especially if you have the patience to sit and silently watch the landscape. Then the coyote may appear in a grassy meadow, searching the grass clumps for unwary field mice. The marmot may whistle from his perch atop a granite rock. The weasel may raise a molelike tunnel with dashing speed in the fluffy, new-fallen snow of a spruce-fir forest. The beaver may leave a V-shaped riffle on the mirror surface of his pond while you wait silently in the aspens above his dam. Count it a day for the record if you see a cougar or mountain lion bounding up a brush-covered rocky slope in the western mountains. They are few in number as are most of the large carnivores. I have seen three of them in forty years of wandering in the wilds.

In the mountains, as elsewhere, there are plant eaters and flesh eaters. This combination assures a system of checks and balances in the environment. If there were no predators, the primary consumers—mice, rabbits, deer, bison, pikas, grasshoppers—could readily eat themselves out of house and home. But within the web of life, energy is cycled from plants to grazers, browsers, seed and fruit eaters, to meat eaters, and scavengers. Following their death the predators and scavengers are also eaten or are recycled by fungi, bacteria, and other decomposer organisms back to the soil again. Thus in a natural community, whether it be a mountain lake, a spruce forest, or an alpine fell field, many animals have functional niches there which combine to perpetuate the living community as a whole.

Even though we do not see a great number of mammals, part of the pleasure in studying wildlife lies in an awareness of their responses to environment, of food chains, and of their importance in ecosystems. Many of these relationships may be surmised by observation of the signs left by their activities. These signs are specific calling cards left by each mammal. When you see them, they tell you that the animal is there. And when the animal is there you have an incentive to study why it is there and what it means to the environment.

The beaver builds its dam to create a pond in which it can build its home and in which it can store its aspen or willow logs for its winter food supply. But since the beaver is there, trout are also there because the pond provides a dependable year-long supply of water. And because the trout are there the kingfisher is there as a part of the food chain—algae, insects, fish, kingfisher, and possibly hawk. The pond water expands the adjoining mountain meadow ecosystem to include grasses and sedges which furnish food and shelter for insects, shrews, meadow mice and ultimately for weasels, coyotes, and horned owls.

In the forest the kitchen midden of pine cone scales left on top of a stump reveals the presence of the Douglas squirrel. Even if you do not see the squirrel himself you know that he clipped the cones by the hundred from the tops of pine trees, collected them from the ground, and shelled out the seeds on his favorite stump or rock. Some of the seeds were eaten and some were cached in hollow logs or buried in soil pits to be dug from beneath the snow for winter food.

The squirrel forgets where some of his cones are hidden. Or the marten, his deadly enemy, catches him and the unused seeds grow into new trees. If there are talus rocks or old woodpecker holes in the vicinity the squirrel sometimes avoids his arboreal enemy the marten. On the ground, however, he is vulnerable to coyotes, foxes, bobcats, goshawks, and horned owls. Thus his seed collecting exposes him to predators and makes him a link in the food chain.

The presence of mammals in a natural community can be determined with greater success if you look for the signs of mammals that live in specific habitats. Deer and elk live in high altitude forests in summer but come down to the lower fringes of the mountains in winter. Porcupines, black bears, red squirrels, lynx, and snowshoe hares prefer the coniferous forests. Big horns, pikas, which are small tailless members of the hare and rabbit family, and marmots live in rocky areas and alpine meadows. Pack rats of different kinds live in different zones ranging from sagebrush-grass communities in the foothills to spruce-fir forests at high altitudes.

By stopping, looking, and listening you may see many of these mammals. You will know where to look if first you see their signs and know they are there. Many a pack rat nest has been located after a camper has missed something from his tent. Once on Grand Mesa in western Colorado a pack rat took eight loaves of bread from my cook kit during the night. A string of bread slices led directly to his nest fifty yards away under a pile of slash left by timber cutters.

Porcupine activity is always indicated by the white patches of wood left on conifer trees where the animal has eaten the bark. If the bare spot is white and still exuding resin it is of recent occurrence. If the scar is dry and healed at the edges it may be several years old and the porcupine may have migrated or

A beaver house sometimes reaches enormous proportions. The house keeps the beavers warm in winter and the ice does not freeze below the underwater entrance.

been killed by fishers, mountain lions, or by foresters. If you know from other signs that a certain animal is present—a beaver pond and dam—you can find footprints of the animal that did the work. A better way to associate animal tracks with their makers is to look at the tracks after you see the animal itself. Few people think to do this but the practice can add immeasurably to one's fund of information about wildlife. Tracks can tell much about an animal's habits and even its sex. A buck mule deer, for example, leaves a track different from that of a doe.

The ungulates or hoofed mammals are among the largest and most conspicuous animals in the mountains. Of these, the bison, bighorn, and moose were formerly more abundant than at present. Unrestricted hunting and unenlightened management has brought about the near extermination of these once truly wild creatures over much of their original ranges. On the other hand, man's activities in cutting forests, killing large predators, and improved game management have resulted in increased populations of deer and elk. The collared peccary or javelina of the Southwestern deserts, foothills, and lower mountain slopes seems to be maintaining its numbers under protection in restricted areas. Formerly, they ranged as far east as southwestern Arkansas.

The tragic history of the bison has often been told. The mighty beasts, in almost incomprehensible numbers, ranged over a large part of the continent. Explorers encountered them on the plains and prairies, along the Potomac River, from western Pennsylvania to the Pecos River in New Mexico, on the desert side of the Sierra Nevada, and in the Canadian wilderness around Lake Athabaska and Great Slave Lake. Daniel Boone commented on the deep buffalo trails leading to Blue Licks in Kentucky. Two races of bison are recognized: the plains bison (*Bison bison bison*) and the wood bison (*Bison bison athabascae*), called the mountain bison by some writers.

The mountain bison ranged up to 12,000 feet in the Rockies. A small herd once lived about Pikes Peak and numerous herds crossed the mountain passes to graze in North, Middle, and South Parks in Colorado. In crossing the Continental Divide these animals went over Mosquito Pass at elevations exceeding 13,000 feet. Apparently they were not bothered by altitude as are domestic cattle. In winter the bison moved down to the valleys to avoid deep snow. The herds survived longer in the northern mountains of Wyoming and Montana, but by 1880 most of the mountain bison were killed by vandals, poachers, and ranchers.

Descendents of the wood bison are now probably hybrids with the plains bison, which were introduced in Yellowstone National Park in 1902 and in Wood Buffalo Park, Alberta, between 1925 and 1929. Bison also are maintained in Wind Cave National Park, South Dakota, and in numerous other preserves and on privately owned ranches. Bison in a mountain environment may be seen in Yellowstone with their calves in spring and summer and sometimes in the Lower Geyser Basin foraging in the snow in winter.

The bighorn, or mountain sheep (*Ovis canadensis*) originally was widespread in the rugged mountains of the western United States. They have been exterminated in much of their former territory but still may be seen in Yellowstone and Glacier National Parks and in Death Valley National Monument. Remnants of the original herds still remain and bighorns are still

The burro, a latecomer to the western mountains, helped tame them. This dependable pack animal carried prospectors' supplies, mining machinery, home furnishings, food, and whiskey over rugged mountain terrain to towns and camps through the West.

hunted in the Rocky Mountains in Colorado, Wyoming, Montana, Idaho, and in Oregon, Washington, British Columbia, and Alberta. Some authorities now believe that trophy hunting for the magnificent horned heads is changing the genetic vitality of the rams with resulting debilitation of the herds and possible final extinction.

Bighorns are among the most sure-footed of all mammals. The sharp cloven hoofs are concave and act as suction cups on the rocks of cliffs, ledges, and mountainsides where the sheep find refuge from predatory animals. They do not spend their entire lives, however, in snow and ice in northern habitats or on sun-parched ledges in desert mountains. In search of food they wander down to gentle slopes where grasses, wildflowers, and shrubs abound. Near Tarryall Peak in the Rocky Mountains I have seen them grazing on the bunchgrass flats at the edge of South Park. In the Southwest desert mountains they come down to the foothills to drink at springs. Usually they appear in bands of half a dozen or more and occasionally are sufficiently tame to allow observation within twenty-five to one hundred yards.

Bighorn rams weigh about 200 pounds at maturity. They are stocky animals with white rump patches and massive curled horns that curl to almost a complete circle. Some of the finest heads have horns that measure nearly four feet along the front curve and have a circumference of fifteen inches at the base. During the rutting season the rams chip and splinter the tips of their horns in violent head-butting battles fought to determine dominance and status in the pecking order of the males. Count it your lucky day if you ever see this age-old ritual of an animal that once lived in extravagant numbers in the mountains of the West.

Another symbol of the wild and high country of the West is the mountain goat (*Oreamnos americanus*). This species, in its nearly inaccessible high mountain retreats, has not been greatly reduced by over-shooting. The goats now occupy essentially the same localities where they lived when the first explorers saw them. Alexander Ross, traveling through western Montana and Idaho in 1823 and 1824, saw them in large numbers. Their populations on the Salmon River have declined as a result of road building in the 1930's and the influx of prospectors and placer miners. Mountain goats still persist in the Rocky Mountains in

Montana and Idaho and from the Cascades in northern Washington northward through British Columbia and into Alaska. Travelers may see them in Mount Rainier, Glacier, and Olympic National Parks and occasionally when driving on roads that cross the high mountain passes in the Northwest.

The mountain goat is about the size of a domestic sheep. Exceptional males may weigh 300 pounds. The female nanny goats are smaller. Long coarse white hair gives the animals a shaggy appearance but the undercoat is thick and woolly. Smooth black pointed horns are produced by both sexes. The black hooves have soft convex pads that enable the animals to cling to rocks and excel in mountaineering in their high altitude summer habitats. In winter they migrate to lower elevations where they feed on grasses, shrubs, lichens, and mosses.

The giant in the deer family is the moose (*Alces alces*). This game animal, the largest mammal with antlers, is possessed of long legs, a body seven feet tall at the shoulders, a huge head with pendulous muzzle, a "bell" of skin that hangs below the throat, and great shovellike flat antlers that spread more than six feet from tip to tip. The moose is generally a solitary animal and does not collect a harem. The great bulls, weighing 1,500 pounds or more, however, engage in mighty battles during the rutting season. Occasionally they lock horns and battle to the death, but mostly they tear up the ground, break down small trees, and end up with gashed bloody hides.

The moose is primarily a Canadian mammal but it ranges southward in the Rocky Mountains through Idaho, western Montana, Wyoming, and northeastern Utah. It also ranges from the northern Adirondacks to northern Minnesota. If you are fortunate you may see this big photogenic animal feeding on water lilies in a mountain lake. It can hold its breath for a minute or more while grazing beneath the surface. You may also see it reaching for leaves and twigs ten feet or more above the ground. Or it may graze on its knees in lush meadows because the long legs make it difficult for the muzzle to reach the ground. I once saw a moose grazing big sagebrush far from the edge of the forest in the Gros Ventre area south of Yellowstone National Park. If you plan to photograph a moose, remember that it is a wary animal and that it may also be belligerent and more than dis-

Wapiti (elk) frequent the high mountain forests in summer but come down to lower elevations when winter snows are deep. They graze in meadows but prefer dense lodgepole pine and spruce timber when danger threatens. In autumn the bulls with mighty antlers fight to retain harems of cow elk.

agreeable. They have driven many a man up a tree and kept him there for hours.

Unlike the moose, which stays with one female at a time, the bull elk (*Cervus canadensis*), monarch of the forest, collects a harem that varies from a dozen to as many as fifty cows. His clear loud bugle in the autumn pines and firs is a challenge to all bulls and a thrilling call that rises to a high pitch, almost a scream, and then drops to a deep grunt. When the challenge is answered by another bull the majestic animals come together with clashing antlers. Shoving, twisting, grunting, tearing the earth with cloven hooves, they battle until one backs away and gallops into the forest. The harem is indifferent as to which bull is victor and which is vanquished.

As the breeding season ends and autumn cold signals the coming of winter, the different bands of elk merge into mighty herds. These migrate from mountain ridges to valleys where their food plants are not all covered with snow. Elk browse on shrubs such as sagebrush, bitterbrush, serviceberry, and aspen. Not until May or June will the spotted calves be born, after the spring migration up the mountains.

The elk, more properly called wapiti, because the true elk is a European animal, is almost moose-sized with huge spreading antlers on males in summer and autumn. The yellowish rump patch, brownish gray body, and white tail distinguish the elk from all other members of the deer family. Originally the wapiti was not only a forest animal but a plains and prairie animal, as were the deer, wolves, and giant grizzly bears. Before white men settled the continent elk were present over much of the United States and parts of southern Canada. Now the Rocky Mountain elk lives in the mountains from New Mexico to western Alberta and eastern British Columbia. The Roosevelt or Olympic elk, a larger and darker variety, lives in the forests of the Coast Range from California to Vancouver Island, British Columbia. Hikers sometimes encounter them along the Hoh River on the trail that leads to the glaciers on Mount Olympus.

The most abundant game animals, known and recognized by almost everyone, are the deer. The whitetail deer (*Odocoileus virginianus*) is the big game animal of the East. Its tail, a large white flag flashing as the animal escapes in the woods, is its distinguishing mark. The mule deer (*O. hemionus*) of the West is the one with the large ears and the black-tipped tail.

The blacktail deer of the coastal mountains from California north to Alaska is now considered to be a variety of the mule deer. Its tail is black on top, the antlers generally are smaller and have fewer points, and its dark coat matches the dimness of the green undergrowth in the dripping redwood and Douglas fir rainforests. Its autumn-spring migrations up and down the mountains are seldom more than a few miles. On the other hand,

Mule deer migrate from high mountains to foothills and deserts for winter food. The fall and spring migrations include thousands of animals in some of the western herds. (Oregon Wildlife Commission)

the mule deer migrate from mountains to sagebrush deserts, sometimes a hundred miles away from their summer ranges.

The fall migration of the mule deer starts when the first heavy snowstorms cover their food supply in the firs, pines, and aspens at 7,000 to 9,000 feet. The bands move down into the sheltered valleys where food is not covered by heavy snow. As winter approaches the bands leisurely move to still lower elevations. In the West their journey may take them one hundred miles or more into the desert to essentially the same territory each year. In late spring they begin their slow trek back to the mountains where they arrive at the higher altitudes late in June.

The fawns are born on the summer range in June or early July. Twins are common and triplets are not unusual. They are weaned by early October and are ready for the migration back to winter range. The bucks, with their new antlers which have been growing since spring, are ready for courtship in late autumn. The males use their horns for fighting with other males but they do not acquire harems of many females as the elk do.

By mid-winter the bucks lose their antlers and become as docile as the does and fawns. The average mule deer is in its prime at the age of six or seven years and is old at ten to twelve years.

A roll call of all the mammals in the mountains reveals that the rodent clan are the most numerous. Most of the order Rodentia, or gnawing mammals, are small to medium size, and must be able to reproduce in great numbers in order to maintain their species. Their world includes many enemies—coyotes, martins, weasels, snakes, predatory birds, and the harsh environment itself. Most of the rodents are herbivores or plant eaters, although some eat insects and other animal food. Each species has a specialized niche within its chosen habitat.

In this niche, life requires response to food, water, snow, temperature, humidity, light or darkness, predators, and disease. Unlike larger mammals which use much energy for maintenance, most of the energy used by rodents goes into reproduction. In their precarious existence few rodents ever reach maturity. Even those that reach adulthood have short life expectancies. In the environmental scheme, rodents are important since they provide sustenance for predators ranging from shrews to weasels, hawks, owls, and even that largest of carnivores, the grizzly bear.

Rodents not only excel in numbers of individuals but their clan includes a remarkable variety. Some are unusual in their habits and in their appearance. For instance, the secretive and seldom seen mountain beaver (*Aplodontia rufa*), which is not really a beaver at all, lives only in the forests along the Pacific coast. It is a cat-sized mammal, with a tail so short it is hardly visible in the animal's fur. The mountain beaver eats fern rootstalks and other vegetation in the Douglas fir forests. In contrast, the ubiquitous deer mouse (*Peromyscus maniculatus*) lives in almost every habitat north of the southeastern states. The beaver (*Castor canadensis*), one of the largest rodents, also once inhabited most of the continent north of Mexico. Trappers nearly exterminated the beavers but they still live in many remote mountain streams and even near towns and cities if they are not molested by people.

The squirrel family includes such a variety of mammals that most everyone is familiar with a few, including chipmunks, prairie dogs, woodchucks, ground squirrels, and tree squirrels.

Of the tree squirrels the most colorful and attractive ones are the tassel-eared squirrels of the ponderosa pine forests on the plateaus above the Grand Canyon. Some authorities recognize two species of this rare animal: the Abert squirrel (*Sciurus aberti*) with tail white beneath, and living on the south side of Grand Canyon; and the Kaibab squirrel (*S. kaibabensis*), with all-white tail, and living on the north side of Grand Canyon. These animals, with the large erect tufted ears, are among the most beautiful rodents on the continent.

One squirrel the visitor in the evergreen forests of the mountainous West is likely to see and hear is the chickaree or Douglas squirrel (*Tamiasciurus douglasii*). This agile chattering sprite is the bane of deer hunters. Seldom do these little dusky colored squirrels with the yellowish bellies and white fringed tails fail to set up a clamor above the deer hunter silently creeping through the woods. The chattering warns every animal in the forest. If the hunter moves along, another squirrel takes up the chatter where the last one left off.

The tree squirrels most of us know are the eastern gray squirrel (*Sciurus carolinensis*) and the eastern fox squirrel (*S. niger*) of the hardwood forests. The gray squirrels not only live in the mountains and the oak-hickory forests of the midwest, but being adaptable creatures, many have become urbanized. You can see them on the lawns in Nebraska City, Nebraska, in the red oak trees in Green Bay, Wisconsin, and in Central Park, New York. In the wild they fill their niche by collecting and planting nuts, building their leafy nests in forest trees, robbing birds' nests, eating the farmer's corn, and raising litters of bushy tailed youngsters. They have planted more trees than all the foresters put together. And they have furnished meat for the stew pot since the first settlers arrived on American soil.

Years ago, George Heinold, in an article in the *Saturday Evening Post*, credited the gray squirrel with helping the colonials be the best marksmen who ever handled a musket. Possibly the expertness gained by our Revolutionary soldiers in "barking off" squirrels helped win the War of Independence. Anyhow, squirrel hunting does improve the eye and the trigger finger of the hunter. In my own time several of my uncles, who were "dead eye" squirrel shooters, introduced me to the art of collecting graybacks in the hardwoods. The cardinal rule was to sit still, and I mean *still*. The second rule was to shoot them in the eye. To spoil the meat with a body shot was a sin.

My uncles also taught me that the ear can be developed for bushy tails. The trick is in learning the thousands of woodland noises made by falling nuts, birds scratching in leaves, branches rubbing one another in the wind, the woodpeckers chiseling on dead wood, the jay's acorn cracking. With experience you learn to distinguish the rasping of squirrel teeth on acorns, the rustle of feet on oak bark, and the low growling of squirrel voices brought on by a dispute within a hollow tree. Sorting out all these noises is one step toward mastery of woodcraft.

Another group of busybodies among the squirrels are the chipmunks. These are ground dwelling squirrels with stripes. They have larger cousins, some with stripes, that also nest in burrows in the ground, among rocks, or in hollow logs. Among these are the marmots of the high western mountains, the white tailed prairie dogs, and the true ground squirrels, of which there are many kinds. But almost everyone recognizes the sprightly chipmunk, so common around campgrounds and vista stops where summer tourists feed the tiny animals with popcorn and peanuts.

The eastern chipmunk (*Tamias striatus*) is the one species found in all of the states east of the Missouri River, northward into southern Canada, and eastward to the Atlantic Coast from Virginia to Maine. Its facial and body stripes, its red rump, and its habit of running with its tail straight up are characteristics. This chipmunk finds timber borderlands more attractive than dense forests. Its tunnel system usually begins under a shrub, beneath logs, in stone piles, or in clay banks. The passageway, about two inches in diameter, slants downward to a depth of a foot or more and then meanders for as much as thirty feet. Somewhere in the tunnel a nest chamber up to a foot in diameter and half a foot or more in height is excavated. Here the young are raised and here an astounding number of nuts and seeds are stored—as many as a half bushel or more. These are eaten in waking periods during the winter hibernation and in early spring when fruits, insects, and other meats are not available.

Unlike the eastern states, the western half of the country has many kinds of chipmunks. Eleven species live in the Pacific States. The least chipmunk (*Eutamias minimus*), the smallest of

the group, is widely distributed and ranges from sagebrush country to altitudes above 10,000 feet. The alpine chipmunk (*E. al-
pinus*) also ranges up to the fell fields and subalpine forests in
the Sierra Nevada. The Townsend chipmunk (*E. townsendii*)
lives in the pine, redwood, and fir forests of California and also
occurs in Mount Rainier and Olympic National Parks. The yellow pine chipmunk (*E. amoenus*) is common in the pine and fir
forests of the Cascades and on the east slope of the Sierra
Nevada.

Other western species include: the cliff chipmunk (*E. dor-
salis*), common in pinyon-juniper forests; the Uinta chipmunk
(*E. umbrinus*), a larger species, found in the Uinta Mountains of
Utah, and near timberline in some of the eastern California
mountains. As its common name implies, the lodgepole chipmunk (*E. speciosus*) in California and far western Nevada lives
mostly in lodgepole pine forests. There are many other chipmunks as various as the habitats and the localities in which they
live. Some of these varieties can be distinguished only by specialists familiar with skull and bone features in museum
specimens.

One beautiful creature that visitors in the western mountains
frequently mistake for a chipmunk is the golden-mantled squirrel (*Citellus lateralis*). This burrowing squirrel, so friendly
around campgrounds, makes its home in the soil near trees,
rocks, or logs. Depending on where it lives—Colorado to California, north to British Columbia, Idaho, and western Montana—
it hibernates from October to late November and emerges from
early March to May. Unlike the chipmunks which go to bed
thin and depend on their stored seeds during waking periods, the
golden-mantled squirrel stores fat to be used as energy during its
long sleep. It is distinguished by its copper colored head which
has no stripes. The white body stripe on each side is bordered by
black.

In the mountains numerous rodents are true specialists. Many
of the mice and voles have such limited home ranges so far from
streams and lakes that they never drink water as do the larger
mammals. Instead, they obtain needed moisture from dew and
from vegetable or animal matter in their diets. Some manufacture metabolic water in their bodies. Other rodents such as the

The golden mantled ground squirrel is one of the friendly rodents that takes peanuts from the hands of tourists in the western mountains. The stripes of chipmunks extend to their heads and this characteristic separates them from the ground squirrels.

marmots, chipmunks, and ground squirrels escape unfavorable seasons by having periods of dormancy or torpor during which their heart and breathing rates are greatly reduced.

One specialist in hibernating is the Uinta ground squirrel (*Citellus armatus*), an inhabitant of mountain sagebrush areas and meadows up to 8,000 feet. Like its cousin, the Richardson ground squirrel (*C. richardsoni*), which lives at elevations up to timberline in western Wyoming and Montana, this rodent sits bolt upright when looking for enemies. This habit gives it the appearance of a stake driven into the ground—from this comes the name "picket pin" ground squirrel.

Shrews exhibit some striking specializations, including plantigrade feet which make them walk with the whole foot on the ground as bears do, long snouts, and a digestive system not matched by any other mammal on earth. The majority of shrews are carnivorous; some eat small amounts of plant food, but the bulk of their diet is insects, spiders, snails, worms, and mice. Some shrews eat the seeds of fir trees. They do not hibernate and since they must eat at least every few hours, their life-long occupation is a constant search for food. Because of heat loss from their tiny bodies and their high metabolic rates they continually face starvation.

Nature has produced a variety of shrews. The dusky shrew (*Sorex obscurus*) lives in wet meadows overgrown with sedges and willows and in coniferous forests where moisture is abundant. It ranges from the Rocky Mountains and the Sierra Nevada north to Alaska. In autumn it stakes out its claim to an area about the size of a city lot by defending its territory against other shrews. If it is lucky enough to find food and escape its enemies it lives through the winter and attains the ripe old age of one year.

The northern water shrew (*Sorex palustris*) is a mouse-sized mammal that runs on the surface of mountain streams or plunges into the depths in search of aquatic insects. Its dark gray back renders it less visible to birds and carnivorous animals from above, and its silvery under surface presumably protects it from large trout and other fish. Thick fur protects it from the chill of icy water in mountain streams.

Possibly the smallest mammal in the world is the pigmy shrew (*Microsorex hoyi*). This diminutive creature weighs from one-seventh to one-twelfth of an ounce—less than a dime. Its tunnels beneath the leaves in wooded areas are hardly bigger than those made by a large insect. Pigmy shrews have been collected in Great Smoky Mountain National Park. They range northward through the Appalachians and Adirondack Mountains and across Canada from Newfoundland to Alaska.

Other shrews that live in various habitats include the least shrew (*Cryptotis parva*) and the shorttail shrew (*Blarina brevicauda*) of eastern United States. The vagrant shrew (*Sorex vagrans*) of the Rockies and Pacific Northwest lives in marshes, along mountain streams, and in other wet places. The masked shrew (*S. cinereus*) ranges across the northern states, Canada, and Alaska in moist habitats and in mountainous regions. Some of the shrews are quite rare. Others are numerous but are seldom seen by the casual traveler in the mountains.

Shrews may be found in meadows and forests by watching for slight disturbances in the grass or among fallen leaves. Although they work mostly below the litter cover on the ground, one occasionally runs across an open space between logs or fallen branches. Their tracks in snow are little blurred patches with tail marks in between. They also tunnel through snow, leaving ridges that resemble the work of miniature moles.

Since shrews have poor eyesight they usually locate their prey by a keen sense of touch. This enables them to find snails, insects, and dead birds, or mice. Living mice, earthworms, and large insects such as grasshoppers when caught by the shorttailed shrew may be paralyzed by poison in the saliva and then stored temporarily in the tunnel system. If these food items are not eaten they add to the organic content of the soil. The tunnels in the leaf litter also aerate the soil and permit moisture penetration. And not least of all, the tiny shrews tends to balance insect and rodent populations, and ultimately to feed larger predators or to return their own body substance to the environment.

The rabbits, hares, and pikas are common mammals in the mountains. Everyone knows one or more of these furry creatures, particularly the cottontail with its big ears, dark eyes, twitching nose, and fuzzy tail. Although jackrabbits are not numerous in the mountains they do live in some localities as high as 11,000 feet. More adaptable in the snowy zones are the snowshoe hares with the big hind feet that enable them to run on soft snow and thereby escape their hereditary enemies, the bobcats, foxes, coyotes, and great horned owls. In the mountains of the West the guinea-piglike pikas are the hay makers of the heights, the one mammal that lives continuously above timberline.

The rabbits, hares, and pikas are rodentlike animals that belong in the order *Lagomorpha*, a group distinguished by two pairs of incisor in the upper jaw—the second pair of teeth lie directly behind the larger pair. The pikas have no tails and their ears are short and rounded. The hares and rabbits have long ears and bushy tails. All are confirmed vegetarians.

Cottontails are represented by several species but distinction is rarely made between them by the average observer. The eastern cottontail (*Sylvilagus floridanus*) lives in brushy areas and open forests from Arizona, Colorado, and the Dakotas to the Atlantic Coast. The New England cottontail (*S. transitionalis*), more reddish in color, is a mountain rabbit of the Adirondacks and the Appalachians. The mountain cottontail (*S. nuttalli*) is the rabbit of the West where it is found from the Rocky Mountains to the Cascades and the Sierra Nevada.

The snowshoe hares (*Lepus americanus*), the ones with the large hind feet and the spreading toes, are truly adapted for life in the snowbound zones in the mountains. In summer they molt to a light brown but in winter they are pure white. They are mostly nocturnal. I first saw them at the Great Basin Experiment Station on the Wasatch Plateau when they come hopping like ghosts on the moonlit snow that was level with the upstairs windows of the cook house-dormitory. They were nibbling on aspen buds and twigs which in summer were ten to twelve feet above the ground. Since then I have seen them in the mountains from New Mexico to Montana and in eastern Oregon and Washington. They are relatively tame and when pursued they circle and return to the vicinity of their starting places.

In spring and fall the snowshoe hare's coat presents a variegated pattern of brown, blackish, and white. Since all the hares do not molt at the same time, many intermediate stages between the summer and winter coats provide an interesting variety of individuals. In autumn the ears, head, and feet first become white. Then the white hairs replace the dark hairs on the back and the rest of the body. In spring the back first changes to brown and the feet are the last to change from white. This change of color with the seasons gives the hares the advantage of concealment even though they do not live in brush heaps or dens.

The interesting little pika (*Ochotona princeps*), sometimes called cony, and a relative of the hares and rabbits, is a true alpine lover. Rock slides and boulder piles above timberline are his habitat. He is a daytime animal and does not hibernate, as does his summer neighbor the marmot. For his winter supply of food he diligently collects grasses, sedges, and wildflowers which are spread on a rock to dry. When this hay is cured it is taken into the "barn" beneath protecting boulders. These haycocks, containing a bushel or more of dried plants, are easily reached by the little animals from the crevices in boulder fields and slide rock where they dwell in winter.

On bright summer days in the high mountains of the West you may see one of the little animals sitting like a bump on a rock. Frequently you hear its squeaky voice before you see the

The pika or cony is active in the alpine zone throughout the year. He makes hay while the sun shines. His dried stack of wildflowers, grasses, and sedges, stored beneath protecting rocks, provides his winter food supply.

pika itself. If you are quiet and patient the little creature will go about its hay making. Quickly it cuts down plants, holds the bundles in its mouth, and dashes over the rocks to the hay pile. It always runs, but occasionally there is a pause—hay in mouth— to look for enemies. Soon it returns for another load. Although the pika is somewhat of a loner, several animals occasionally collect hay in the same vicinity near a communal rock pile. Each, however, seems to have a home territory which it defends against other pikas.

The pika has few mammal neighbors in winter. When the snow lies deep on alpine slopes the deer and elk move down to more congenial valleys and foothills. The mountain goats and bighorn sheep seek lower altitudes on sunny slopes. The pocket gophers remain active below ground, constructing burrows in search of root food and pushing up earth cores in tunnels made in the snow. Meadow voles use the gopher tunnels in moist meadows. The marmot has long since gone into his winter sleep. Only the pikas and their weasel enemies remain in the boulder fields and rock slides, the one eating from his pile of hay, the other incessantly hunting for gophers, voles, ptarmagin, and pikas.

Each natural community has its own combination of plants, plant eaters, carnivores, scavengers, and decomposer organisms. Many of the larger mammals use more than one community during the cycle of the seasons. Elk and deer move from high summer ranges to the foothills or even to sagebrush deserts. With them go the larger predators, cougars, wolves, and coyotes. The largest predators, the bears, however, tend to remain in the forested zones at middle elevations where they eat a great variety of vegetable matter during the summer and where they go into a long sleep in their dens in winter.

The grizzly bear (*Ursus horribilis*), king of the woods, is a vanishing species. These great hump-shouldered, dish-faced, hairy monsters with the four-inch claws once roamed the mountains from New Mexico northward into Canada and Alaska. Before white men came they ranged far out on the plains and prairies in search of big meat, including elk, deer, and bison. Now they are restricted to remote wilderness areas in Wyoming, Montana, and Canada. Because of the danger to humans there is much controversy concerning management of grizzlies in Yellowstone and Glacier National Parks where they still may be seen by tourists.

The black bear (*Ursus americanus*) still is abundant and is an important game animal in the mountains of the western states. In the East it prefers the forests in the Appalachians and the timbered swamps in New England. Recently I saw tourists feeding a black bear on the road to Clingmans Dome in Great Smoky Mountain National Park. The greatest concentration of black bears, however, seems to be in the state of Washington where the bear population has been estimated to be as high as 50,000. Several years ago I killed a bear on Temperance Creek which flows into the Snake River in Hells Canyon in northeastern Oregon. Bear signs on the trails indicated a probable population of one or more bears per square mile in this isolated mountain wilderness.

Bears are where you find them, and you can find them in many places. Recently, one wandered down the foothills of the Rockies into the city of Fort Collins, Colorado. I still chuckle about the black bear I saw while deer hunting below Windy Point east of Mount Hood, Oregon. I shot twice to scare him and he bounded up the slope, over the hill and right into Bill

The black bear lives in the eastern and western mountains and is common in Yellowstone National Park. The population in Washington state has been estimated at 50,000.

Courrier's camp. There the bear dismantled Bill's tent, scattered the groceries, and ate the bacon and eggs. Instead of being scared the bear demonstrated his omnivorous appetite.

A bear will eat almost anything in the way of food: berries, chipmunks, honey, carrion, crippled game, such as deer or elk, ants and larvae from rotten logs, grass, and acorns. A huckleberry patch in Oregon or a serviceberry grove in Colorado is a likely place for finding a bear. Occasionally a bear will go on a rampage in a sheep herd at night, striking right and left with powerful forepaws until a dozen or more victims lie dead.

One summer, when I was doing range research on the Uncompahgre Plateau in western Colorado bears were doing damage to the sheep. The herder constructed log enclosures with traps inside baited with dead sheep. Apparently the guilty bears were caught because the damage stopped. The thing I remember best about the episode is how we rendered out a gallon of bear grease for use in waterproofing our boots. The stuff was so penetrating it made my engineer's boots wrinkle down around my ankles like loose stockings. No one had told me to mix the bear grease with beeswax and neat's-foot oil!

Black bears are largely nocturnal but they also wander about in the daytime. Few people see them because the bear generally avoids a person unless the two happen to meet suddenly on a trail. In spite of the animal's size—300 to 500 pounds for a mature bear—it can move quickly and silently through shrubs and trees. Injuries from people-bear encounters are few and far between in the natural environment. Elimination of garbage disposal pits and the ban on feeding bears in the National Parks has markedly reduced bear injuries to tourists.

The mountain lion (*Felis concolor*), also called cougar, panther, painter, catamount, deer tiger, puma, and a host of other names is a beast of mystery. This cat with the long slinking body and heavy cylindrical tail has long been a legend because of the innumerable stories about its secretiveness, its prowess as a hunter, its voice, its attacks on people, and its depredations on domestic stock. There is substance to many of these tales, but the romance of this striking animal is vanishing under the pressure of hunting and the invasion of the wilderness by modern civilization.

Originally, the mountain lion ranged over most of the North American continent. Gradually it was extirpated in the eastern states; the last panther in the Great Smoky Mountains, for example, was reported to have been killed about 1920. In 1882 my grandfather recorded in his diary that he threw sticks of firewood at a "catamount" that was disturbing the chickens. This was at his home on the edge of the prairie a few miles north of where Rock Creek flows into the Little Nemaha in southeastern Nebraska. About that time a few mountain lions were seen on the Niobrara and Loup Rivers in Nebraska. Now the range of the great cats is mostly limited to the mountainous areas of the West where deer furnish their principal food.

The numbers of mountain lions have been greatly reduced by consistent trapping and hunting. Bounties paid by the states have attracted some of the world's most dedicated lion hunters; California, for example, paid bounties on 10,558 lions from 1907 to 1950. One of the most famous of California hunters, Jay C. Bruce, killed 581 lions during the 28 years he worked for the State Division of Fish and Game. C. W. Ledshaw, the second state lion hunter, turned in 342 lions for bounty. In the Southwest, Ben Lilly, last of the mountain men, served as a government hunter and killed so many lions by tireless pursuit with his puma dogs that he became a legend.

For many years there has been controversy whether a mountain lion screams or not. Mountain lions spit and growl menacingly. Many have been heard to yowl in zoos as would any large cat. Lion kittens make a wheezy meow. But there seems to be no record of anyone ever seeing a mountain lion while it was screaming like a child or a woman in agony. Jack E. Handy, who formerly lived in Durango, Colorado, and paid his college expenses with lion bounties from the Denver Post, later experimented with lion voices while hunting near Twisp, Washington. He and a doctor friend dissected the voice boxes from mountain lions and blew air through them with bellows. Jack told me he could produce all manner of screeches and other sounds by this method. His record of kills exceeded one hundred.

When mountain lions formerly were abundant their impact on the natural food chain must have been considerable. Competent scientists estimate that a mature lion will kill at least one deer each week during the winter season. When lion populations amounted to hundreds or even thousands in a single western state one can easily calculate the control exerted on deer herds.

The bobcat is an efficient predator. His diet includes birds, rodents, rabbits, bighorns, and even an occasional small deer.

The impact also extended to mountain sheep, elk, beavers, peccaries, and porcupines. Other prey taken by mountain lions include marten, marmots, pack rats, bobcats, mice, and wild turkeys. Livestock killed by mountain lions include horses, steers, colts, calves, sheep, burrows, goats, and pigs.

Another efficient predator that ranges high into the mountains is the bobcat (*Lynx rufus*). This cat with the short ear tufts and the black-tipped tail has a body and head length of two to two and one half feet. The tail is less than six inches long. The buff colored body and white under parts are black spotted. Bobcats have been known to kill young deer and bighorn sheep but their usual prey consists of small rodents and birds such as quail and grouse. The animal is nocturnal and is seldom seen unless hunted and treed by dogs.

The coyote (*Canis latrans*) is one of the most maligned and persecuted mammals in America. Stockmen accuse them of every form of depredation on their livestock, hunters shoot them on sight, government trappers poison them, and the environmentalists appeal to Congress to outlaw the use of nonspecific poisons which not only kill coyotes but also other animals and beneficial birds, including hawks and eagles.

The coyote does kill sheep and even calves, especially when prairie dogs and other rodents have been poisoned by overzealous self-appointed experts who forget that rodent populations frequently increase because of livestock overgrazing and mismanagement of range lands. But in spite of the persecution the coyote still persists. If you do not see him in the daytime you may know that he is still around when his *yap, yap,* followed by a chorus of barks, wails, and howls echoes across the dark landscape.

Coyotes will eat almost anything. They function in the ecosystem as predators on grasshoppers, ground squirrels, pocket gophers, mice, rabbits and frogs. They also eat porcupines, crayfish, carrion, garbage, pine needles, juniper berries, apples, plums, and watermelons. They can distinguish between ripe and green watermelons. In pursuit of jackrabbits they work in pairs or family groups. There are numerous reports of how coyotes accompany badgers. The badger digs out rodents and if he misses, the coyote catches the prey.

A multitude of insects—bees, butterflies, moths, flies, and beetles—
pollinate wildflowers. Blister beetles, shown here, commonly occur on
goldenrods and other composites. The larvae of some of these beetles
feed on grasshopper eggs.

8

A Multitude of Lesser Animals

OF ALL THE ANIMALS in mountain forests, meadows, and even on alpine summits, the insects and other arthropods are represented by more species than any other group of living things. At times some of these invertebrates appear in overwhelming numbers and cause awesome destruction of forests and shrublands. Thousands of acres of trees sometimes are destroyed by bark beetles, cone moths, nut weevils, sap-sucking insects, and defoliators of young trees. Numerous insects are injurious to mountain forage plants grazed by deer, elk, rabbits, mice, and other herbivorous mammals.

At the same time, multitudes of other insects are so useful that the mountain environment would be barren of vegetation without their presence. Many are flower pollinators, some of which are so specialized they pollinate only certain genera or species of plants. Small bees of the genus *Osmia*, for example, are dependent on pollen and nectar from certain groups of *Penstemon*. On the other hand, some *Penstemon* flowers are so structured that the bees do a better job of pollinating than do other insects. Furthermore, this business of pollinating that goes on before our eyes may be more remarkable than one might first suspect. The *Osmia* bees also visit flowers of raspberries and clover. The questions arise, Do the larvae, which live in cells packed with leaf segments, require a mixture of pollens for proper nutrient balance? What effect does this selective pollination by one group of insects have on the genetic development of *Penstemon?* These, and a thousand other questions can be raised for insects wherever they occur.

The insect population in mountain environments is an essential part of the whole complex of living, reproducing, and dying. Without insects, fungi, bacteria, and innumerable other invertebrates, a mountain forest soon would be choked with the debris of dead leaves, branches, and tree trunks that fall as a result of lightning, fire, flood, and old age. The reduction of this debris to organic compounds, duff on the forest floor, and ultimately to soil, makes room for new growth and thus sustains the whole cycle of life.

Even though foresters bemoan the loss of timber resulting from insect attack—losses that are sometimes catastrophic—nature in the long run restores the landscape and renews the living mantle of the earth. Insects play an important part in this restoration through conversion of living and non-living material into substances that become usable again by other animals and other plants.

A roster of insects found in mountain environments includes virtually all the orders of insects. These live in an incredible diversity of habitats and niches. Some, such as the chewing lice, feed on feathers of birds or the skin of animals. Many are parasites on other insects. The springtails are primitive, wingless insects equipped with a springlike device, appended to the fourth segment of their body, which enables them to jump several inches. They are found on snow, on water, in moist soil, under bark, and on fruit and dead animals. Other insects live in water. Fishermen know these, particularly the dobsonflies because their larvae, the hellgrammites, make excellent fish bait. They also know the Mayflies which cause trout to rise in great numbers when a "hatch" is on and the adults appear in countless numbers. Their nymphs live on stream bottoms under stones and among algae and other submerged plants.

The conspicuous insects are the butterflies, moths, and skippers. These belong to the second largest order, the Lepidoptera. Several thousand species occur in the United States and many are found in the mountains. As a group they are well known to everyone because of their colorful wings and bodies, and because of the scales or flattened hairs which rub off like dust on one's fingers when they are handled. They are further distinguished by the mouth parts which are coiled up like watch springs. But it is the incredible beauty of some of the butterflies and moths that makes them the best loved of insects.

Butterflies and moths both have brilliant colors but most moths conceal their markings by folding their wings flat while at rest. Butterflies commonly fold their wings upright. They are diurnal insects whereas the moths fly mostly at night. Butterfly antennae are knobbed at the tips. One Superfamily, the skippers, have antennae hooked at the tips. Most moths have plumose or feathery antennae.

It is somewhat of a paradox that the adults perform the useful function of pollinating flowers, while their larvae are voracious feeders, sometimes on the very plants their parents pollinate. Sometimes they destroy their own food supply but seldom to the point that both consumer and host vanish from the scene. As is usual in nature, predators and diseases come along to balance out their numbers and the living community returns to an approximation of its normal structure.

The larvae or caterpillars of butterflies and moths are all similar in structure; some are brightly colored; and some are even ferocious in appearance. The regal moth (*Citheronia regalis*), for example, produces a caterpillar known as the hickory horned devil, which reaches a length of four to five inches. When disturbed it raises its head and front end with reddish black-tipped horns in a startling display of fierceness. However, it is completely harmless. It feeds on a variety of trees, including walnut, hickory, ash, persimmon, butternut, sycamore, sourwood, and sumac.

Some of the large moths of the eastern forests are endowed with incredible beauty. Most people are familiar with the cecropia moth (*Hyalophora cecropia*) which has a wingspread of about six inches. The green head and body, the featherlike antennae, and the wings with white crescent-shaped spots and red-bordered cross-bands make this giant insect a most dramatic creature in the woods. Other species of amazing beauty are the promethea moth (*Callosamia promethea*) with its reddish purple to brown wings, the luna moth (*Actias luna*) with the delicate green wings, and the polyphemus moth (*Antheraea*

Tent caterpillars of many kinds infest trees and shrubs in the mountains. The adult insects do relatively little damage. (U.S. Forest Service)

polyphemus), brownish-yellow with a transparent eyelike spot on each wing. The larvae of these great moths spin large tough cocoons attached to branches, enclosed in leaves, or in debris on the ground. The larvae feed on many species of trees including ash, birch, cherry, walnut, sassafras, hickory, and sycamore.

Moths of smaller size are exceedingly abundant and are variable in their markings, habits, and food preferences. The adults are mostly nocturnal but many are attracted to lights at night. Among the common ones are the tussock moths with tufts of hairs on some of the segments of the larvae. The fall webworm is a widely distributed species. Its larvae spin silken webs which sometimes cover small trees. They are not selective in their diet since they eat the leaves of more than one hundred species of trees. The poplar tent maker (*Ichthyura inclusa*), common in New England, constructs webs on poplar and willow trees by drawing the edges of leaves together with silk.

The geometrid moths are a numerous clan of more than 1,000 species and are interesting because their larvae travel by forming a loop and then extending the body forward while holding on with their rear legs. Because of this method of locomotion they are known as measuring worms, loopers, inch-worms, canker worms, and spanworms. When we were children and one of these larvae crawled on us, someone was sure to say, "He's measuring you for your coffin." The larvae of the geometrid moths are common in forests where they feed on many kinds of trees.

Moths and butterflies are not abundant in alpine regions. Some of the arctic species, however, do live near and above timberline. One of the mountain butterflies, the melissa arctic (*Oeneis melissa semidea*), also called the White Mountain butterfly, is restricted to the alpine regions of the White Mountains in New Hampshire. A race of the widely distributed polixenes arctic (*O. polixenes Katahdin*) occurs on Mount Katahdin in Maine. Some of the other far northern alpine butterflies range southward in the mountains but occur at elevations below timberline. One of these is the common alpine (*Erebia epipsodea*), found in meadows in the Rocky Mountains south into New Mexico.

In contrast with these local races of butterflies, some species are cosmopolitan because of their migratory habits. The painted lady (*Vanessa cardui*) migrates from its overwintering places south of the Mexican border in massed flights of countless millions. The width of the advance may be one hundred miles and the direction may not always be northward. A great southward flight occurred across Wyoming in 1965.

A northward flight in 1973 east of the Rocky Mountains lasted for a week. I could count up to thirty individuals passing through our front yard south of Fort Collins at any given moment. The migration also extended into the foothills and up into the ponderosa pine zone of the Rockies. Reports from Boulder, Colorado, stated that there were so many butterflies, they gave the appearance of thin clouds moving a few yards above the ground. A few of the painted ladies deposited eggs on the thistles in the fields near our home and I also had to hand pick the larvae from the hollyhocks in our garden. But the mass migration flew northward and I never learned where they came to rest.

The California tortoise-shell butterfly (*Nymphalis californica*) sometimes appears in large numbers in the western states. Ordinarily their larvae feed on ceanothus shrubs. Several years ago I came upon millions of them hatching in the ponderosa pine woods in the Cascade Mountains west of Wenatchee, Washington. The butterflies with brown wings below and orange above, black bordered and dotted with marginal rows of purple spots on the hind wings, were resting everywhere on bare ground. But more interesting were the pendulous chrysalids dangling by the hundreds from every redstem ceanothus shrub on the mountainside. When I touched a bush the chrysalids vibrated on their attachments to the twigs, setting up a rattling that soon was transmitted to other shrubs. I have never seen this before or since, and I have no explanation for how it was done or what purpose it served.

In the mountains there is always an abundance of hymenopterans—ants, bees, and wasps. The presence of some of these is made conspicuous by their works. In other instances the insects themselves are distressingly present. The hymenopterans, including some of the ants, can sting viciously.

On numerous occasions I have seen riders bucked high wide and handsome when their horses were being stung by hornets after brushing one of the large paper nests. Although it was dangerous business, the men on one of our range research crews on the Wasatch Plateau in Utah used to take great delight in stirring up a hornet's nest behind an unsuspecting worker. The villain would poke the nest with a stick and then lie face down without moving in the grass or weeds. The victim would suddenly yell and then run like a wild man through the woods or crash through bushes like a bull elk gone beserk. The stings were painful and sometimes the victim's eyes would swell shut.

The yellowjackets are hot-tempered wasps that build their paper nests in the ground. The abdomens of these insects are banded with black and yellow and to me have always seemed to have a cameolike beauty. They are especially fond of nectar and

Yellowjackets are common in western mountains. They are fond of
sweet fruits, and around camps will eat jelly, fresh meat, and fresh fish.

ripe fruit juices and in the presence of sweets are exemplary but
pestiferous companions. On the Uncompahgre Plateau our research crew learned to eat jelly sandwiches with care. The yellowjackets would eat on one end of our sandwiches while we
ate from the other end. The process involved gentle waving with
the right hand and taking a quick bite after making sure no hornet was in the spot where we wanted to bite. It was impossible
to keep the insects away but no one ever was stung while eating
a jelly sandwich.

The cynipid wasps are among the most accomplished of gall
insects. The females lay eggs on twigs and leaves of trees, particularly oaks, and the larvae produce irritating substances which
stimulate the tree to produce galls or swellings of the plant tissues. These structures come in many shapes and colors. Some
are like apples, others resemble small blisters on leaves, and still
others are small hemispherical structures covered with white
hairs. Stem galls sometimes enlarge until they resemble sweet
potatoes. A few are shaped like limpet shells or cone shaped tents
and others are shaped like tiny goblets.

The great majority of galls are not particularly injurious to
the plants on which they occur. Infested branches do become
disfigured and occasionally they die, or even the entire tree is

killed. One species of gall wasps produces galls in acorns which may reduce the seed crop. Sweet substances produced by the galls attract bees, birds, and ants. The gall wasps also are parasitized by chalcid flies, thus adding another link to the energy chain.

The ants are among the most numerous insects in the mountain environment. One species or another can be found all the way from the foothills to the alpine zone. Some are small and live in tiny colonies of only a few dozen individuals in a shallow hole in the ground or beneath a protecting stone. Others, such as the carpenter ants, are large and perform mighty labors by chewing holes in dead trees and reducing woody debris to sawdust that returns to earth and replenishes the fertility of the soil.

Ants are social insects that live in communist societies. They practice division of labor but every task is done for the good of the community as a whole. They are intelligent, to a degree, but they are bound by instinct to tasks so amazing as to be almost beyond belief. Their jobs in the colony include food gathering, nest construction, eating, sleeping, baby sitting, caring for their larvae, defending their home territory, and burying the dead. Ants also change jobs from day to day. An individual worker may dig tunnels one day, husk seeds on another day, collect honeydew from aphids on sunny afternoons, and transfer eggs, larvae, and young ants to warmer or cooler parts of the nest as the temperature changes from hour to hour or from day to day.

Some species of ants are seed gatherers; others sip nectar from flowers; and many are hunters of worms, insects, and other kinds of meat for the family larder. A large nest of black *Formica* ants that I studied in the ponderosa pine forest in eastern Oregon contained an estimated third of a million ants. The dome-shaped earthen mound was nearly ten feet in diameter at the base and nearly three feet high. On warm days thousands of these ants explored the woods around their home and climbed grass stems and leaves. On close examination I learned that they were collecting Pacific grass bugs which are detrimental to forage grasses. In addition to doing a good turn by protecting the grass for grazing animals the ants were collecting protein for the colony food supply.

In the same forest the large nests of thatching ants were conspicuous. These ant mounds were fabulous collections of twigs,

A wood ant (*Formica*) nest built against a stump in the ponderosa pine country in Montana. A full sized twig nest may contain 10,000 to 200,000 ants.

pine and fir needles, pieces of bark, and other debris collected from the forest floor. The largest nest was thirty-four inches high and seventy-eight inches wide at the base. Some of the nests had pits a foot or more in depth below the ground level. These pits also were filled with twigs, many of them larger than kitchen matches, portions of dried leaves, and remains of beetles and other insects. A few of the nests were constructed over tree stumps which added to the permanance of the mounds since these colonies persist for many years.

One of these large nests may hold up to a half million ants. From the mother nest, trails that are veritable Appian Ways lead to small suburban ground nests where queens and small numbers of workers are maintained. Here new colonies can start if the main nest with its numerous egg-laying queens is destroyed by forest fires or other accidents.

Many trails lead to the hinterlands, sometimes 200 feet from the nest, to trees and shrubs where the ants tend aphids for their honeydew and collect insects and plant parts for food and construction materials. The highways back to the nest are jammed with homeward bound ants with gasters swollen with liquid food or mandibles clutching caterpillars, bird dung, and other trophies of the chase.

Hornets, yellowjackets and other wasps are natures' papermakers.
Some hornet nests are a foot or more in diameter. The entrance hole is
at the bottom of the waterproof paper nest.

All these materials are useful in the life of the colony. In the
nest, dozens of queens are busy laying eggs; hundreds of workers carry out the dead; thousands more make alterations and repairs in the nest; and ten thousand care for the eggs and feed the
grubs and young ants. Obviously, a notable amount of energy is
exchanged in an ant city of this size. And certainly millions of
ants have an impact on the forest environment and on the food
chain that involves more than the insect world.

There are many architects and artisans in the mountain insect
world. Some build nests of clay and stone; others make houses of
paper, manufactured from wood of forest trees; and many construct tents and tentlike shelters of silk made within their own
bodies. Most of the large nests are community projects built by
the social insects—wasps, hornets, ants, and termites. Other insects make dwelling places for individuals by rolling leaves or
sewing twigs, pine needles, or even debris together with silk to
make log cabins, bags, and egg cases.

The nests made by the papermaking wasps are elaborate and
wonderfully constructed affairs. These globular homes of the
hornets, sometimes two feet or more in diameter, are fabricated
from dead and weathered wood by chewing and mixing the

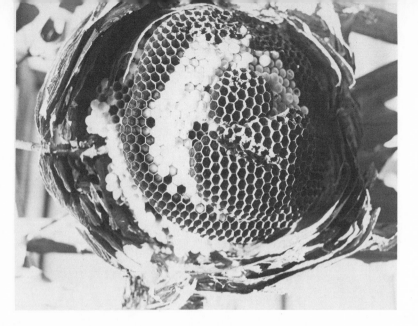

Internal design of a hornet nest resembles the honeycomb made by
bees. The cells that contain eggs and larvae, however, are made of paper
instead of wax. The cells hang downward and do not become
waterlogged even if rain penetrates the nest through a rent in the
paper cover.

wood fibers with a salivary secretion which results in a kind of
paper. The wasp adds the moistened pulp to a gradually growing
sheet that develops in layers as multiple walls are formed in the
nest.

The nest first begins with a stem or pedicel attached to a tree
limb or other support. Then brood cells are constructed for the
first eggs to be laid by the queen. Then paper walls are added to
enclose the brood and an entrance hole is left at the bottom of
the nest. Ultimately several tiers of brood comb are constructed
as the nest is enlarged and remodeled inside. The outer walls of
paper are separated by air spaces which insulate the interior
against sudden temperature changes from day to night. New
queens and drones are not produced until late in the season; in
late autumn all the wasps die, excepting the new queens which
seek sheltered places for hibernation over winter. This end to
the colony is also true of the yellowjackets which build their
paper nests and combs in cavities in the earth.

While the wasp and bee tribes are famous for building paper
or wax nests, there are other insects skilled in complicated construction work. The termites fabricate aerial tunnels over sticks

and stones to reach woody material that serves as food. These tubes, cemented together with particles of soil, enable the termites to work and travel without exposure to the air. Ants, on the other hand, travel in the open but some of their thatched houses are marvels of construction for temperature control, air conditioning, and protection from rain and snow. The carpenter ants do not eat wood as the termites do, but they excavate large galleries and intricate tunnels into the heartwood of fallen logs while making their nests. The discarded sawdust is blown away and eventually becomes a part of the forest soil.

The decomposer organisms in forest litter and wood are linked together in a chain of bacteria, fungi, and insects that play an essential part in recycling the elements necessary for continuation of life. When a tree dies and falls to earth the recycling process has already begun. Woodpecker holes, bark beetle engravings, and ant and termite galleries have permitted entry of bacteria and mycelia of mushrooms and other fungi. As the disintegration process continues, a succession of other insects, pill bugs, worms, millipeds, and other animals add to the process of dissolution until the wood finally crumbles away and its particles in the soil finally contribute their substance to the continuing world of living organisms.

The long-horned or roundheaded borers are among the most numerous and accomplished wood-boring beetles in the forest. More than 1,400 species are known in the United States. The larvae are fleshy, thin skinned, white or yellowish in color, and are possessed of efficient mandibles. They girdle twigs and stems, chew bark, bore in roots, and excavate galleries in the wood of dying or fallen trees. Some are twig pruners and others specialize on cured lumber and structural timber in houses.

The adult long-horned beetles are robust insects and many exhibit spots or bands in various color patterns. They are distinguished by their oblong bodies, long legs, and strong flying ability. Many have wide spreading antennae that are two or three times longer than the beetle itself. Some make squeaking noises when captured. The larvae of the genus *Monochamus* are known as "sawyers" because of the loud noise they make while feeding. A stethoscope is not necessary to hear the grinding, crunching, rasping, and clicking sounds that arise from a log where these two-inch long borers are chewing through solid wood.

Carpenter ants do not eat wood. They have excavated their nest in this Douglas fir log and deposited the sawdust outside where it will be blown away by the wind and become a part of the forest soil.

The work of insect artisans in leaves, bark, and wood produces some of the most graphic designs in nature. The patterns left by leaf skeletonizers, leaf miners, gall makers, sawyers, girdlers, and borers often are intricate and unique in appearance. The biramous, or many branched galleries, of the native elm bark beetle sometimes show a definite symmetry. The beetle bores a hole through the bark, constructs a nearly horizontal gallery, and lays eggs close together on both sides of the gallery. When the larvae hatch they bore away from the gallery, some up and some down, following the grain of the wood. The egg galleries of the pine engraver (*Ips grandicollis*) also run longitudinally, grooving both bark and wood. The larval mines are more or less transverse and the tunnels get larger as the larvae bore through the wood and get larger.

The ambrosia beetles (pin-hole borers) and the Columbia timber beetle have special structures, called mycetangia, which store and transport fungus spores. When the spores are introduced into trees and logs they produce fungi which stain the galleries black or brown. These fungi are used by adults and larvae as food. Thus the insects and fungi form a microsymbiotic relationship. Each species of beetle has its own specific ambrosial fungus, and trees are selected for attack largely on the basis of the growth requirements of the fungus. Some of the larvae of boring beetles remain in wood for two or three years before they reach adulthood. In that time they can riddle a tree and bring about its death. Whether we detest the depredations of these insects, or are fascinated by their ingenious solutions to the problems of existence, we must admit that they are successful animals and that they are intimately woven into the web of life.

From the sequoia woods in California to the forested mountains of Appalachia some of the most intriguing animals are the frogs and salamanders. They are oddities from the standpoint of structure and appearance, because of their life cycles and their evolutionary history.

The amphibians were the first animals with backbones to crawl out of the ancient seas and begin life on land. This happened some 325 million years ago. Their emergence from water required a number of adaptations for survival. Lungs were necessary for air breathing so they could obtain oxygen. They had to get rid of tails and fins and had to develop legs for locomotion. And finally they needed hearing, sight, and smell adapted to life on land in order to perceive enemies, locate mates, and find food.

They are fascinating animals because they still retain some of their ancient features and because they are still land and water animals. The frogs and toads, even those that live in prairies and deserts, still lay their eggs in water and their tadpoles live in water until they absorb their tail, and metamorphose into adults. The salamanders, lizardlike in form, also lay jelly-covered eggs, usually in water, that develop into larvae with external gills. Gradually the larvae transform into adults.

Frogs and toads can hear by means of an eardrum or tympanum, a circular membrane located below and behind the eye. Studies indicate they hear sound frequencies only about half as high as we can. Salamanders lack hearing by means of eardrums but can detect surface vibrations through their front legs or lower jaws, depending on the species.

The courtship of some of the salamanders includes a form of dance, with caresses between the sexes, figure eights, waddles, and neck rubbing. This puts the female in the proper disposition to pick up the spermatophore deposited on the ground by the male. She inserts this sperm in her body and the eggs are fertilized internally before they are laid.

Salamanders occur in streams, ponds, and moist places in the prairie, in the desert, and in the mountains. The southern Appalachians have long been regarded as the home of a large variety of amphibians. These range from the hellbender, two feet or more in length, the fifth largest salamander in the world, to very small species such as the pigmy salamander which is seldom two inches in total length.

Some salamanders, such as the mudpuppy, are well known. The Blue Ridge Mountain salamander is common in moist habitats to the summits of the highest mountains. The cave salamander and Wehrle's salamander are found inside limestone caves and in moist ravines. Altogether there are possibly forty or more species of the so-called lungless salamanders in the Appalachians. The adults of these breathe through their skins and the linings of their mouths since they have no gills or lungs.

The variety of salamander species and races in the southern Appalachians is due to isolation on mountain peaks through thousands of years rather than adaptation to different habitats. During the Ice Age, forests with habitat conditions suitable for salamanders were distributed throughout the lowlands as well as in the higher mountains. When the glaciers retreated northward and the climate warmed, spruce forests, for example, persisted on the highlands but vanished from the valleys, leaving the salamanders no routes of migration from one mountain group to another. Through time the salamanders on isolated mountain masses developed specific racial characteristics.

The Appalachian salamanders come in many varieties and colors. The green salamander (*Aneides aeneus*) occurs from West Virginia to Alabama and is distinctive because it is green. The red-cheeked salamander (*Plethodon jordani jordani*) has cherry-red cheek patches and a blue-black body. The dusky salamander

Tiger salamanders inhabit ponds and lakes in the mountains. The greatest variety of salamanders occurs in the southern Appalachians where various species and races are isolated in different mountain habitats.

(*Desmognathus fuscus*) is a variable fellow: some specimens are blackish or brownish, others are almost yellow, and some are reddish. The spotted salamander (*Ambystoma maculatum*), a relative of the tiger salamander of the West, is marked by yellow or orange spots on a gray or slate-colored background. It breeds in winter in the southern Appalachians.

Salamanders are not so numerous in the Rocky Mountains, partly owing to more severe climates, drying winds, and forests not so uniformly moist as those in the Appalachians. One species, the tiger salamander, is more abundant than most people realize. A friend recently told me he had never seen a salamander in the mountains. Then he drained the fish pond on his ranch in the Front Range of the Rockies. "The bottom of the pond," he said, "was literally covered with salamanders."

Tiger salamanders, with yellowish or white patches on gray backs and sides, are widely distributed through the United States. They live in moist places in the prairies and in ponds at 10,000 feet elevation in the Rockies. They live in seclusion during the dry season, in damp basements of houses and in burrows made by prairie dogs, badgers, marmots, and ground squirrels.

At least twenty kinds of salamanders live in the far western states. These include the newts, which are essentially aquatic, the mole salamanders which stay underground most of their lives, and the lungless salamanders. Some of these are restricted in distribution. The Jemez Mountains salamander, for example, is found only in the Jemez Mountains near Los Alamos, New Mexico, at altitudes up to 8,750 feet. The Pacific giant salamander, on the other hand, ranges from California to British Columbia in humid coastal forests and inland to the Cascade Range in Oregon and Washington. Another species of similar distribution is the Escholtz salamander for which several subspecies are recognized. They appear in numbers when the breeding season begins in February and March.

Although most people are not familiar with these secretive creatures, the salamanders are worthy of contemplation. Being of ancient lineage they are reminders of the tree of life wherein these amphibians gave rise to reptiles which in turn branched into the bird and mammal groups. But somehow the salamanders also survived for millions of years.

The frogs and toads are a numerous clan over much of the continent, but we are inclined to think of them as inhabitants of ponds, streams, and marshes in the lowlands. Many species, however, live in mountainous country where water, land, and even trees, enable them to live their double lives as tadpoles and adults. Wherever water is found, even in alpine habitats, you may see these amphibians.

The peep of tree frogs in eastern mountain forests is one of the first signs of spring. These little animals are expert climbers and one may expect to find them and their relatives clinging to twigs or leaves high in trees. The voices of different species are so distinctive that identification is possible by sound alone. Frogs, unlike birds that sing to warn intruders away from their territories, use their voices to attract female frogs for mating.

The northern spring peeper makes a single piping note in late winter and early spring. The eastern gray treefrog produces a harsh trill. This is the one with an X mark on its back. The swamp cricket frog, a terrestial species, produces a metallic sound which has been likened to the noise of a finger pulled over the teeth of a pocket comb. This frog lives in swamp lands, marshes, and meadows. It has been found at 11,000 feet in the Uinta Mountains in Utah. The chorus frog's voice is heard in early summer in the mountains in Yellowstone National Park. Common frogs in Appalachia include the green frog (*Rana clamitans*), the pickerel frog (*R. palustris*), and the wood frog (*R. sylvatica*). These are inhabitants of streams and ponds, generally at elevations below 4,000 feet, since suitable waters are rare at higher altitudes. The wood frog makes an early appearance in the southern mountains; breeding may start in early January, but is more common in February. Wood frogs occasionally are found in limestone caves.

Frogs have a multitude of enemies. Their tadpoles are eaten by wading birds, fish, giant waterbugs, dragonfly larvae, diving beetles, and by other tadpoles. The adults are eaten by large salamanders, crows, herons, skunks, raccoons, opossums, and snakes. The pickerel frog, however, turns the tables on snakes. It contains a strong poison sufficiently virulent to kill a snake within half an hour after the frog is eaten. Snakes have learned to avoid pickerel frogs in favor of the similar leopard frogs (*Rana pipiens*) which are harmless.

Toads also range widely in the United States and some are found at high altitudes. This is not strange when we perceive that many insects which serve as food for amphibians also occur even in the alpine zone. Grasshoppers, flies, mosquitoes, and spiders are especially abundant in mountains, even near their summits. Many of these are apparently blown from lower elevations by high winds. Grasshoppers breed at high elevations and are active during the day when temperatures are favorable. There are no crickets in alpine country, probably because their nocturnal activities are not adapted to the low night temperatures on mountain tops. The toads, however, do not go hungry since they adapt to a wide variety of food.

The western toad (*Bufo boreas*) is widely distributed from Alaska to California, Idaho, Montana, and Colorado. Its habitats include beaver ponds and clear streams on mountains and plateaus up to 10,000 feet. Bees, beetles, ants, grasshoppers, spiders, and even crayfish are a part of its diet. Its voice is a clear birdlike chirp.

The Yosemite toad (*Bufo canorus*) occurs in the high Sierra Nevada of California at elevations ranging from 6,500 to 10,000 feet. It breeds in pools, streams, and the cold water from melting snow. On land it hides beneath stones, logs, and in rodent burrows. Its voice is a long trill with ventriloquial quality which makes it hard to find even when it is singing.

There are many other frogs and toads in the mountain environment. We do not know all that is to be known about them; even the breeding activities of some have never been observed. Some are unique, especially the American bell toad or "tailed frog" (*Ascaphus truei*), which lives in the cold mountain streams and forests along the Pacific coast in Oregon and Washington. Its distribution also includes southern British Columbia and parts of Idaho and Montana.

This unusual frog has some of the characteristics of salamanders: short ribs attached to some of the vertebrae; a tongue that does not protrude to catch prey, as in other frogs; no tympanum, which means that it cannot hear as do other frogs; and no voice. In the rushing mountain streams the voices of other frogs probably could not be heard anyway.

The most interesting feature of this frog is the "tail," which is not a tail, but an organ for copulation possessed only by the male. This organ permits fertilization of the eggs inside the female, which assures greater success than would be possible by external fertilization in swift rushing water. The female attaches the eggs to rocks in the srteam bed and when the tadpoles hatch they cling to the rocks by lip suction, even though they are able to swim in the fast moving current. Truly, the tailed frog is an animal admirably adapted to a demanding mountain environment.

Magpie babies remain together when they first come out of the nest.
They develop long tails as they grow older.

9

Birds of the Mountains

THE FASCINATING VARIETY OF BIRDS in the mountains derives from the many environments in the altitudinal zones beginning in the foothills and ending in alpine meadows. Each zone has its distinctive climate, seasons, vegetation cover, and variety of habitats. Consequently, many kinds of birds are able to find congenial living conditions for their individual and specific life requirements. Even within elevational zones there are microenvironments suitable for ground birds, bush birds, tree birds such as jays, phoebes, and gnatcatchers. In the tree tops and above are crows, hawks, and vultures. On the cliffs are swallows, falcons, and eagles.

Birds are more numerous in the lower elevation zones. The hardwood forests of the Great Smoky Mountains, for example, with their oaks, hickories, red maple, sourwood, and other trees and their understory of mountain laurel and rosebay rhododendron, provide habitats for the tufted titmouse, whippoorwill, eastern phoebe, brown thrasher, chipping sparrow, and more than a dozen kinds of warblers. Higher up, in the northern hardwoods, dominated by beech, yellow birch, maples, mountain silverbell, and cucumber trees, are homes for the solitary vireo, scarlet tanager, and the yellow-bellied sapsucker.

The spruce-fir forest above the hardwood zone is less congenial for birds where the limited number of tree species provide less variation in habitat. Among the red spruce and Frasier fir trees and in the openings occupied by pin cherries, mountain ash, and yellow birch you may find the dark-eyed junco, black-capped chickadee, and blackburnian warbler. In the grass balds, or high altitude meadowlands you may see wild turkeys, ruffed grouse and the ruby-throated hummingbird. Above Clingmans Dome you should watch for duck hawks and common ravens.

In the Rockies and other high western mountains only a few birds are content to live in the cold changeable weather of the alpine country high above timberline. Strangely, it seems, the water pipit, horned lark, and rosy finch live here in summer and find an abundance of insects and spiders blown up from lower levels by the wind and cooled to inactivity on the snow fields. Hummingbirds attracted by the alpine flower gardens also seem out of context with the environment at these cold high altitudes. Their tiny bodies seem unsuited to the cold. The wild turkey, with its huge bulk in relation to its surface area, would appear to be better adapted than the tiny mites with their whirring wings. Metabolism must be part of the answer for the hummingbird. And the motor that produces the rapid wingbeats must keep him warm, as does the motor in some of the moths and night flying insects.

Other mountain birds, at first glance, also seem to be out of context with their environments. The ptarmigan, for example, resides on windy slopes among the mountain peaks, exposed to wintery blasts, drifting snow, and enemies that run the gamut from hawks, eagles, foxes, martins, to mountain lions. They, and some of their grouse relatives, move so deliberately and apparently so fearlessly that you almost have to push them aside with your feet while walking across the tundra. But they can move quickly and explosively when necessity demands. When you see their feathered feet, concealing coloration, and diet of arctic willow leaves, you can understand how they survive as resident birds among all the vicissitudes of one of the most demanding environments on earth.

One of the fascinating aspects of bird study in mountain habitats is not identification or compilation of life lists, but the search for the meaning of the presence of birds. This involves knowledge of their habitats, food and shelter requirements, breeding cycles, and the environments in which they live. Observation of even the common birds affords opportunity for speculation about their adaptations to such things as the food supply and to one another.

The nuthatch, for example, walks down the trunks of trees while the brown creeper goes up. Possibly this gives these birds different angles of view in their search for insect prey in the same microenvironment. Hence, their niches, or ways of life, are different and they avoid competition with one another, at least in a small way. On the same tree the chickadee examines twigs, the wren searches leaves, and the crossbill opens cones. The black-and-white warbler, however, acts as a combination creeper-nuthatch. This small bird of the foothills and mountains moves up, down, and around tree trunks. But its slender bill and aerial ability also enable it to sip nectar, eat fruits, and catch insects on the wing. None of these competes with the ground birds searching for beetles, spiders, worms, or fallen seeds, or with the woodpecker excavating ants and insect larvae from dead branches in the tree top.

A pleasing variety of birds occurs in most of the mountain habitats, especially in the hardwood and pine zones. Birds are likely to be abundant in the vicinity of lakes, rivers, and in forest openings where the transition zone between trees and grasslands provides greater variety of plant and insect food and a wider choice of nesting sites. Rocky canyons frequently support bird species not found in other mountain habitats. In the canyons on the east side of the Cascade Mountains, for example, you may expect to see violet-green swallows soaring and diving beneath your perch on the canyon rim. In the desert mountains of Arizona the Abert towhee, Bewick's wren, and the Arizona pyrrhuloxia are common residents along with possibly such rare wanderers from Mexico as the coppery-tailed trogon. In winter, you might even find a hibernating Nuttall's poorwill clinging to the rocks in a canyon wall.

Along the dashing streams in the remote wilderness of the West it is always a pleasure to encounter birds that seem to be out of place. I am inclined to think of great blue herons as denizens of fresh water lakes in the Great Plains, of marshes on Cape Cod, of the swamps in Florida, or of the salt flats of Hood Canal in Washington. They nest, however, in the tall Douglas firs along Tumalo Creek in the Cascades of central Oregon. And they fish in the high mountain lakes in the Colorado Rocky

Rufous hummingbird babies in nest decorated with lichens. These hummers build their nests in the dark understory of fir and hemlock forests from western Oregon to western Canada.

Mountains. They adapt to almost any habitat if fish, frogs, and other foods are available.

Franklin's gulls, so common on the prairies of the United States and Canada, sometimes appear in the mountainous country of Wyoming along the Snake River and the Yellowstone, hawking the caddis flies like swallows while the trout are jumping from the waters below. The shy and colorful harlequin ducks normally summer along arctic shores. But they provide an occasion to be recorded when you see them being tossed in the icy cataracts of rivers in the Northern Rockies of Montana. They seem to enjoy the challenge of the foaming rapids in a world of jagged snow-covered mountain summits and dark coniferous forests.

In contrast, the smaller birds frequenting the lower mountain zones from Maine and New Hampshire to Tennessee and Georgia provide the avian enthusiast with an endless variety of species to see, hear, and study in all seasons. Many of these are just plain old songbirds that most of us know and love. From the

The gray jay is the darling of the camp in coniferous woods. It is very
tame and will steal bread or bacon from the skillet on your picnic table.

Adirondacks to the Great Smoky Mountains the plethora of
common birds include the dark-eyed junco, white-throated sparrow, yellow-bellied flycatcher, Canada jay, hermit thrush, purple finch, and various warblers.

The avifauna of mountains, because of elevational and vegetation variations, provide a rich variety of bird affinities where
northern and southern, eastern and western, bird distributions
overlap. In the southern Appalachians, for example, northern
affinities include the ruffed grouse, saw-whet owl, veery, and
Blackburnian warbler. Along with these are common residents
such as the pileated woodpecker, belted kingfisher, blue jay, cardinal, and the largest bird of all, the turkey. Summer residents
include the catbird, wood thrush, summer tanager, and the ruby-

throated hummingbird, the smallest bird in the eastern United States.

A different combination of avian affinities intrigues the bird student in the Black Hills in South Dakota. Here, numerous species reach the eastern edge of their breeding range: western tanager, western wood pewee, Lewis' woodpecker, Audubon warbler, and many other western birds. In this same territory are common eastern species: house wren, ovenbird, and American redstart. Ruffed grouse drum in the ponderosa pine woods. Continent-wide species, including the gray jay, the red crossbill, the kinglets, and the golden eagles mingle with the multigeographic assemblage.

Of all the insect eating birds the woodpeckers and their kin seem to be specialized the most. Their ability to extract insects and larvae from solid wood and their nest drilling activities make them unique among woods-dwelling birds. The cavities made by woodpeckers in trees, stumps, and even in clay banks are sufficiently distinctive to enable one to identify the species in most instances. In addition, the woodpeckers add drumming to their voice calls for communication purposes. Their interactions with other birds and mammals exert definite influences on the forest environment.

Woodpeckers and flickers have been among my favorite birds since my boyhood days in the oak, elm, cottonwood, and basswood forests near the Missouri River in southeastern Nebraska. In memory, I associate their presence with bright colors and high visibility on sunny days from January to April, when the young leaves begin to appear on the Dutchman's breeches and make bright mats of flowers on the moist forest floor. Then the drumming of the red-headed woodpeckers on whitened snags in the dead treetops and their piercing cries in flight are at their best. More subdued in the open deciduous woods were the *wick-a, wick-a, wick-a,* notes of the common flickers—notes which sounded to me like corks being twisted back and forth in the necks of bottles. The first time I heard the hissing sound of the young in a dead tree trunk, I thought it might be snakes. Then when the parent birds fed their fledglings a fight in the hole sounded like a scuffle between barnyard cats.

I learned early that flickers, or yellow-hammers, as the local farmers called them, explored the ground somewhat in the manner of robins searching for worms. Woodpeckers still are among the most vivid bird acquaintances one can make.

The giantic pileated woodpeckers, big as crows, and sort of clumsy clowns of the woods because of their size, are the most striking of them all. The drilling, chopping, and chiseling activity of these birds is a spectacle to behold. As extractors of large carpenter ants from dead standing and fallen timber these birds have no equals. When they are at work, bark splinters and chips, some several inches in diameter, fly in all directions or accumulate in a trash pile at the base of the tree. They can bore a hole half a foot wide and six inches deep in deadwood and leave vertical furrows several feet long. The opening into the nest cavity may be seven or eight inches across and the cavity itself may be two feet deep.

The pileated woodpeckers function as more than insect exterminators in the woods. The cavities these birds dig become homes for owls, chickadees, wrens, bluebirds, mice, and even tree-hole mosquitoes when water accumulates in these excavations. These woodpeckers add to the impact of other birds on insect life in the forest by biologically protecting trees. But they do not keep all trees from dying. Lightning, insects, and fungus diseases kill some trees in the natural environment. And formerly, before they were persecuted by foresters, porcupines did their share of work in providing suitable trees for the wood chiseling birds and the subsequent guests in their abandoned nest cavities.

The acorn woodpeckers of the California foot-hill oak and pine community are specialists at fitting acorns into holes drilled for the purpose. These gregarious birds with the red caps, black chins, dark backs, and white wing patches, live among the valley oaks, blue oaks, coast live oaks, and their numerous other kin, and the digger pines, mostly at elevations below 4,000 feet. They are so assiduous at digging pits in bark, fence posts, and telephone poles, and pounding acorns into these holes that even the wood matrix sometimes is almost concealed. The birds do little damage since they seldom penetrate to the cambium layer beneath the bark of living trees.

The acorn woodpeckers, however, are not completely addicted to nuts. They drill for larvae in dead wood and even chase

Dead ponderosa pines furnish homes for woodpeckers and other insect-eating birds. Insects and fungi in fallen trees return the logs to their natural elements and thus enrich the soil.

insects on the wing. Unlike other woodpeckers, they tend to nest in colonies where eggs and young are tended without great regard for individual ownership. Obviously, the acorns stored so conspicuously in wooden pits also are communal property.

Two woodpeckers of the lower mountain slopes bordering the desert, are the Gila woodpecker and Mearn's golden flicker. These birds chisel nesting holes in the saguaro cactus. When the cavities are abandoned they are used by the Arizona crested fly-catcher, the screech owl, the elf owl, and lizards, mice, insects, and spiders. Thus the woodpeckers in arid lands are also linked in a chain of interdependence for food and shelter.

Throughout more northern habitats woodpeckers are notably

present in most seasons of the year. Many of these are birds most of us know. The red-headed woodpecker with its bright red head and neck and plain white breast cannot be mistaken for any other bird when seen in the open deciduous woods. In addition to its usual diet of wood-dwelling insects and their larvae, the woodpecker includes acorns in its winter fare. When the crop of beech nuts is plentiful it commonly remains all winter in the northern Appalachians.

The downy woodpecker is one of the commonest of the eastern woodpeckers. It also occurs in deciduous woods in the West. Living in similar habitats the hairy woodpecker, somewhat larger than the downy, can be recognized by its long bill and the white stripe down its back. In the oak covered mountain slopes in California, Lewis' woodpecker is attractive because of its pink front, red face, and greenish back plumage.

The yellow-bellied sapsucker, widely distributed throughout the United States, is another woodpecker with singular habits. Its predilection for drilling parallel rows of small holes in live trees serves a double function for the bird. The holes produce sap on which the birds feed. The sap also attracts insects which the birds eat when they return to the scene of their depredations. Extensive drilling through the cambium layer sometimes kills the tree.

In the rugged terrain of the western mountains the northern three-toed woodpecker is resident in the thick forests above 7,000 feet. This bird with the black wings, barred sides and back, and yellow cap is a secretive creature and is seldom seen. It is rare in the East. In the Rocky Mountains it commonly nests in dead lodgepole pines. It aids in control of destructive bark beetles which become numerous in spruce stands that have been blown down by heavy winds.

The predatory and scavenger birds of the mountains and high mesas are a varied lot and generally of large size. Some are fish-eating birds of prey, including the osprey, the bald eagle, and wading birds such as the great blue herons which frequent mountain lakes and streams in the pine and fir zones. White pelicans are found on lakes in the Coast Ranges along the Pacific shore. On Yellowstone Lake in northwest Wyoming these huge white-and-black birds are summer residents with breeding colonies on the Molly Islands. Here they rely on cutthroat trout for food.

Golden eagles, falcons, crows, and ravens also occur in the mountain wilderness where they afford much bird watching pleasure since they often fly at great heights and are highly visible above the forests and rocky pinnacles. More secretive and common on the western plains, deserts, and mountains are the magpies, master scavengers and quite as smart as crows in their depredations on small animal life. They are diurnal birds, as are the turkey buzzards which are the most accomplished scavengers of all. Owls of various kinds also inhabit the mountains, but one must be alert to see them since they are mostly nocturnal birds of prey. As a group, the predatory and scavenger birds perform important functions in maintaining the environmental balance of living creatures. Many of them stand near the top of the food chains in their chosen habitats.

Today it is a rare sight when a bald eagle soars into view on its seven or eight foot wings, plunges to water to capture a fish, or with majestic diving power intercepts an osprey laboring up to its eyrie with a fish. The eagle snatches the fish out of the air and carries it to its own young in a gigantic stick nest on a towering tree or a niche in the rocks on the mountainside. Sometimes, when the eagle itself catches a huge spawning salmon too heavy to carry, the eagle's claws become embedded in the back of the fish and cannot be disengaged. Then the bird is carried under water and drowned.

Bald eagles and golden eagles occur in the northeastern mountains and are seen by hawk watchers in the Appalachians. In the West, the bald eagle is rare where it once was common. Guns, traps, and environmental poisons including DDT have reduced their numbers almost to the vanishing point. They still may be seen in the Yellowstone country where they are protected in the National Park which once was an ideal habitat for this symbol of our nation.

The golden eagles are dark brown and not truly golden in color. A pale brownish cast over the nape feathers and back has given them their name. Unlike bald eagles they seldom nest in trees; instead, they establish their eyries on sheer cliffs, rock ledges, or even mud banks of creeks. In spite of their majestic appearance in flight and the eagle's reputation for prowess and valor the golden eagle will desert her nest and young without a fight when a human intruder appears. Golden eagles occur throughout the United States at one season or another. They are typical birds of deserts, grasslands, and mountains.

The eagle-sized osprey, or fish hawk, is a wide ranging bird of seashores, lakes, and rivers, since it feeds entirely on fishes. Many are seen on the spring and fall migrations both in the East and in the West. They are easily distinguished from eagles since they are dark above and clear white below. A dark wrist patch also marks the crook in each wing.

Visitors at the falls in the Grand Canyon of the Yellowstone see ospreys at their eyries on the rock columns above the rainbow spray between the yellow canyon walls. Their stick and driftwood nests also are built in tall trees that are broken off and dead. The nests are used year after year. Tree nests of ospreys also are numerous in the dead trees in Crane Prairie Osprey Management Area, Deschutes National Forest, in the central Oregon Cascades. The birds there are protected by the U.S. Forest Service and the population in 1971 was 200 birds. The birds seem willing to use artificial nesting platforms which are being built as old snags rot and drop into the reservoir.

Unlike the golden eagle, the osprey will defend its nest against eagles, hawks, and human intruders. In its fishing habits there is no more accomplished fisherman than the osprey. When it "stoops" like a falling rocket there is a great splash. Then the bird shakes the water from its feathers as it flaps to gain altitude with the prey clutched in its talons. It also captures fish near the surface by plunging only its feet into the water. Like the eagles its numbers are declining from the effects of guns and pesticide pollution.

Among the raptores the numerous species of hawks are both maligned and admired. Many hawks are killed because of their methods of obtaining food, especially if their prey happens to be chickens in the barnyard or ruffed grouse which hunters want for themselves. Some of the falcons also kill ducks, but their predation actually makes little impact on the duck population as a whole.

The hawks and falcons are a varied clan. Some are rare, especially the goshawk and the gyrfalcon. These are birds of the Arctic tundra and the far northern forests. The diet of the gyrfalcon on the tundra includes ducks, ptarmigans, lemming, and Arctic hares. When the food supply fails in the north country they sometimes wander into the northern states. Then their plunder of grouse, ducks, and even herons can be severe. Happily, their stay is usually short and because of their esthetic value they are tolerated, especially by bird lovers.

Hawks and falcons are commonly seen in timbered areas in mountainous country. In the West some of the frequently observed species include the sharp-shinned hawk, rough-legged hawk, Swainson hawk, prairie falcon, pigeon hawk, and sparrow hawk. Species usually found in the East include the red-tailed hawk, broad-winged hawk, pigeon hawk, Cooper's hawk, and the perigrine falcon. The autumn migrations of these birds over some of the eastern mountains are so spectacular that raptor enthusiasts sometimes plan their trips to the major lookouts a year in advance.

Hawk Mountain, Pennsylvania, is the most famous of the lookouts. Above the Appalachian Valley is the Kittatinny ridge where hawks and eagles on migration rise on updrafts of air and sail past the promontory which affords the finest view of hawks on the continent. Because of the height of the ridge it is possible to look down on many of the passing birds.

The autumn migration usually starts in September and continues through October. Hundreds of birds may pass in a single day. Cool temperatures and steady northwest winds are likely to bring the best flights. One record flight occurred on September 16, 1948, when 11,392 hawks were counted. A good day for broad-winged hawks may result in a count of more than 2,000.

September is the best month for broad-winged hawks. Ospreys may appear in numbers for a few days along with lesser numbers of sharp-shinned hawks and sparrow hawks. Marsh hawks in small numbers appear all through the migration period. The best flights of sharp-shinned hawks and red-tailed hawks pass by in October when the flaming vegetation colors are at their height. Golden eagles, goshawks, and sparrow hawks may also be expected. October also brings migrations of rough-legged hawks, pigeon hawks, peregrine falcons, Canada geese, vireos, thrushes, and wood warblers. The spring migrations are not so spectacular since the birds travel northward at a more leisurely pace, and they take various routes between and over the Appalachian ridges.

Although Hawk Mountain Sanctuary offers the most famous autumn spectacle, other worthwhile spots are scattered throughout the Northeast. Donald S. Heintzelman in *A Guide to Northeastern Hawk Watching* lists numerous hawk lookouts, including some along the Atlantic seashore. These include Mount Cadillac in Maine, Mount Belknap in New Hampshire, and Mount Tom in Massachusetts. The latter presents a display of hawks almost equal to that of Hawk Mountain in Pennsylvania. Mount Peter in Orange County, New York is a good autumn lookout. In West Virginia, Hanging Rocks Fire Tower, on Peters Mountain near Waiteville, is an important spot for autumn observations.

There are no hawk lookouts in the West that compare with those in the East. One might expect to see hawk migrations along Steens Mountain and over the Owyhee Mountains in southeastern Oregon but this area is so remote that few observers interested in hawks ever visit these mountains. Each year multitudes of waterfowl wing up and down through Blitzen Valley, that portion of the Pacific Flyway between the Jackass Mountains and Steens Mountain, a long high ridge east of the village of French Glen, at the southern tip of the Malheur National Wildlife Refuge.

Hawks nesting here locally include the red-tailed hawk, Swainson's hawk, marsh hawk, ferruginous hawk, prairie falcon, sparrow hawk, and the golden eagle. In the high grass and sagebrush covered mountains that rise above Trout Creek east of the hamlet of Fields, Oregon, these and the sharp-shinned, Cooper's, and the rough-legged hawk may be seen, at least occasionally. If there is a marked autunm migration over these mountains, I do not know; but it seems to be a likely territory.

The year-round resident birds in the mountains are a varied lot. In spite of their differences in size and choice of habitats, they are adapted to the vicissitudes of the seasons, especially the cold of winter. The one thing common to all seems to be that the chosen and sometimes varied diet of each species is available at all seasons. Because of this the majestic turkey, the lackadaisical grouse, the mischievous black-billed magpie, and the diminutive dipper are able to flourish in environments inhospitable to many migratory birds. Some of the resident birds, such as the white-tailed ptarmigan, remain above timberline throughout the year. Others migrate to lower elevations when the winter food supply becomes critical at high altitudes.

The native gallinaceous or chickenlike birds, once widely distributed throughout North America, still occupy a variety of habitats in the mountains. Among these, grouse are interesting because of their strange behavior and the bizarre courtship of some of the species. The prairie chickens, the first grouse with which I became acquainted, used to display on their booming grounds on my uncle's ranch in western Nebraska. The resonant calls of the males, their display of orange air sacs, the fanlike tails, and the clash of wings during their fights, made them unique among all the birds I had known on the prairie. These, of course, were not mountain birds. But my boyhood acquaintance with them made a fitting introduction for observation of the habits of other grouse I was to see in later years.

The ruffed grouse, an inhabitant of rugged mountain terrain, conifer forests, open fields and apple orchards, occurs far down the Appalachians, across the forested northern part of the continent and southward in the Rocky Mountains and in the mountains of the Pacific Northwest. The ruff on the breast, the barred tail with the terminal black band and the bird's habit of drumming by beating the air with its wings distinguish it from the other grouse.

The ruffed grouse is a challenge to hunters. I know, because I have blasted off many a piece of bark with my shotgun just as the grouse flew behind a tree trunk. In spring and summer in the Blue Mountains in eastern Oregon, over a period of ten years, I learned the location of several dozen "drumming logs" in the pine and fir woods where the ruffed grouse put on their displays and drummed in both daylight and in darkness. Most of these logs were under the heavy cover of spruce trees, undoubtedly a protection from flying predators.

These grouse do have many enemies, especially the great horned owls, and the snowy owls when they come down from the Arctic regions in winter. Weasels catch some of the adults as do bob cats, foxes, and coyotes. The eggs and young are taken by skunks, raccoons, opossums, and squirrels. Even the chipmunk occasionally steals the eggs. The chicks survive partly through camouflage and through their astounding immobility among weeds and grasses when danger threatens. The adults avoid many enemies by roosting in spruce trees or by diving into soft snow in winter and sleeping under its white insulating cover.

Food for the ruffed grouse and the seldom seen spruce grouse of the northern coniferous forests is hardly a problem. The young birds include insects in their diet. But the adults apparently eat a variety of herbaceous materials including wintergreen berries, acorns, beechnuts, pine needles, aspen buds, teaberry leaves, and fern tips. In winter they eat a variety of buds from shrubs and trees.

In contrast, the sage grouse, the largest of the American grouse, is essentially an eater of sagebrush leaves, particularly in winter. The sage grouse summers in foothill country and winters on sagebrush plains and deserts. It is not specifically a mountain bird although thousands of them live in the high mountain parks in northern Colorado at elevations up to 10,000 feet. In the sagebrush covered slopes in the Aspen Creek and Pueblo Mountains in southeastern Oregon I have seen dozens of them around my camp in aspen groves and on the mountainsides at elevations above 8,000 feet. In this remote wilderness of sagebrush-grassland they are hardly molested by man and consequently are almost tame.

The tamest of the grouse is the spruce grouse. The subspecies, known as the Franklin grouse in the mountains of northeastern Oregon, the Cascades in Washington, and the mountains of British Columbia, hardly moves even when sticks are thrown at them. In the Wallowa Mountains and in the northern Cascades I have had them walk between my feet while I was writing notes on range vegetation. They well deserve the name "fool hen."

The blue grouse of the western mountain and coastal coniferous forests, unlike most birds, migrates upward in winter to dense evergreen forests. Eight geographic races of this bird are recognized, including the dusky, Richardson's, Swarth's, Oregon, Sitka, and Sierra grouse. One distinguishing characteristic is the orange comb over each eye. The hollow hooting sound made by the male always reminds me of the sound of the pump

The blue grouse lives in open woods in summer and migrates upward to mountain fir forests in winter. Males have a yellow or orange patch of skin above the eye. These birds are tame and will hardly move aside when you meet them in the woods.

on the kitchen sink in our Nebraska farm house. At other times it sounds like an old fashioned gasoline engine just starting up. The call is disconcerting since it seems to come from different directions, making it difficult to locate the bird even when it is near at hand.

At times these birds seem to be quite fearless. When flushed they sometimes sit in a nearby tree and watch, even if you throw stones at them. Recently I encountered one on the road that had been plowed from ten-foot snow drifts on Mt. Lassen in California. When I attempted to photograph it the bird came forward and battered my camera with its stout wings. When I tried to hold the camera with my right hand and push the bird back with my left hand it slashed my left index finger with its bill so that blood gushed forth. I'll always remember being "grouse bit" on Mt. Lassen.

During the seven years when H. E. Schwan and I periodically did range research in the alpine country from New Mexico to Montana we occasionally encountered white-tailed ptarmigan. I believe that for every one we saw a thousand saw us. Usually we saw them singly or in pairs in summer. In winter we saw

Magpie young are always hungry and their resourceful parents feed them grasshoppers, pieces of dead mice, young birds, and animals killed by predators or by automobiles. The adults are jay-sized birds with long greenish-black iridescent tails.

them in their white plumage in groups of a dozen or more. Formerly they could be observed from the highway over Vail Pass in central Colorado and around Engineer Mountain in the San Juan Mountains. These were birds of the southern race of white-tailed ptarmigan (*Lapogus leucurus altipetens*), in the alpine meadows from northern New Mexico to northwestern Wyoming and south central Montana. The other races live in the mountains of western United States, Canada, and southern Alaska.

White-tailed ptarmigans are unwary birds. They depend largely on camouflage for safety. In winter they blend so completely with the snow that only their slow movements reveal their presence. The black bill and dark eyes are the visible marks by which one might recognize them. During the molting season they exhibit a mixture of whites, blacks, and browns. In summer the wings, tail, and belly of the male are white. The mottled plumage of the females blends perfectly with the lichen-covered rocks where they hide their nests. The nests are so camouflaged on the wind-blown tundra that few people ever see one.

The opposite extreme in nests is that of the black-billed magpie, who builds a nest that is both a monstrosity and a work of art. This black and white jay-sized bird with the long and seemingly unmanageable tail is at home on the western plains and deserts and on the lower mountain slopes. During mild winters these birds may be seen in the ponderosa pine zone from Colorado to Idaho and eastern Oregon.

Magpies are expert scavengers and have received notoriety because of their alleged depredations on livestock. Zebulon Pike, the discoverer of Pikes Peak in Colorado, accused them of pecking holes in the backs of his pack mules when food was scarce in winter. Numerous other reports tell of magpies pecking in the backs of sheep and stealing ducklings and young chickens from farm yards. When I used to hunt ducks in Colorado I did not dare leave my game unattended when magpies were present. These self-sufficient birds, however, are noted for cleaning up dead animals and are expert grasshopper catchers.

The magpie nest proper resembles a large robin nest. Instead of mud, the magpie sometimes uses cow manure to cement together the fibers, grass, plant stems, and twigs that comprise the true nest. Around this sturdy bowl a tangled mess of sticks is inserted in the bush or tree until the whole affair is the size of a bushel basket or even larger. The sticks point in every direction, both above and below and around the nest, to make an impenetrable fortress against large climbing mammals and hawks and owls. Usually an entrance is made on one side but occasionally an escape hole is made on the other side of the bowl. The largest stick nest I ever saw measured approximately two feet in diameter and six feet in height.

One more resident bird that travelers see and enjoy in the western mountains, from western Mexico to Alaska, is the dipper or water ouzel. This drab short-tailed creature, about the size of some of the larger sparrows, is common along mountain streams where it is truly adapted to a watery environment. In summer or winter it stands on a rock, bobbing its body up and down, and then calmly walks into and under the water in search of aquatic insects and other food. It emerges completely dry and sings with melodious notes. It builds its nest among moss covered rocks, sometimes behind a waterfall, where the spray constantly moistens both eggs and young. It is one of the most dependable birds I know for showing visitors. If you know where

Magpie nests are conspicuous in juniper trees in the foothills of the Rockies and Cascades. These twig structures enclose the true nest which is shaped like a very large robin's nest. The entrance hole in this nest is near the tree trunk.

a dipper lives, sooner or later it will appear. And it will bend its knees and bob or dip every second or two.

In the Southwest numerous mountains rise above the desert and the desert grasslands. These mountains support a variety of vegetation zones with habitats suitable for birds not adapted for life in the surrounding desert. Many of these mountains are rugged, some are dry and harsh, but others exhibit spectacular beauty spots amid cool forests. Shaded canyons open into green valleys and parklike pine country. This mosaic of rocky peaks, high mesas, deep chasms, flower-covered slopes, and thick forests offers great variety of habitats for resident birds and for migrants that come north for the summer or fly south to escape the harshness of northern winters.

Some of the southwestern mountains are high. Wheeler Peak, near Taos, New Mexico, reaches 13,161 feet above sea level. Some of the mountains along the Mexican border are dry and desertlike to their summits. Others are timbered. All present a colorful diversity of land forms, especially the Magdalenas, the

Clark's nutcracker is a fearless visitor in mountain campgrounds. With an appetite like its cousins, the jays, crows and magpies, it eats bacon at your camp table and takes peanuts from your hand while it is on the wing. The light gray body, black wings and white patches on wings and tail are distinctive.

San Mateos, the Ladrones, and the Animas Mountains. In Arizona, many of the mountains rise abruptly above the desert and some achieve heights that support spruce, fir, and aspen stands typical of forests on mountains hundreds of miles to the north. Among the enchanting mountain ranges are the Huachuca Mountains near Douglas, Arizona, the Catalina Mountains near Tucson, and the Chiricahua Mountains in southeastern Arizona.

Many of the birds in these desert mountains are old friends we see elsewhere, but are of different races. The western bluebird, for example, breeds from southern British Columbia, Idaho, and western Montana to southern California. The variety, *bairdi*, called the chestnut-backed bluebird by some authors, is resident from Colorado and Utah south to the Mexican highlands. It is one of the colorful birds among the pine and firs in the Chiricahua, Hulapai, Santa Catalina, and Huachuca Mountains in Arizona and the Guadalupe Mountains in Texas.

In the coniferous tree zones and mountain parks the great variety of birds ranges from the tiny broad-tailed hummingbird with its baffling gyrations among the showy meadow flowers, to the majestic Merriam wild turkey, which may weigh close to thirty pounds. The abundant mourning doves nest from the plains clear up in the coniferous forests along with their close relatives the band-tailed pigeons. These pigeons are extremely adaptable. I have watched them from the Front Range of the Rockies in Colorado to the Oregon shore of the Pacific Ocean. Near Portland, they come down daily from the Douglas fir forests south of Mt. Hood to the salt springs along the Pudding River. When collected during the hunting season their crops still contain berries that grow on the slopes of Mt. Hood and in the high Cascades near the source of the Clackamas River. The round trip of fifty to one hundred miles between the mountains and the soda springs is a daily routine for these birds.

The raven, black and distinctly larger than a crow, is a common resident bird from the foothills of the California Coast Ranges to the highlands of Texas and northward through the deserts and mountains to Washington, Oregon, and Montana. In Texas you may see them from the Rio Grande to the highest elevations in the Chisos Mountains. These accomplished aviators fly from their roosting places in the mountains to feed each morning in the desert. The ravens are scavengers that compete with magpies for rabbits killed on highways and for parts of deer and other mammals left by hunters. Along the Trail Ridge Road in Rocky Mountain National Park ravens become tame enough to compete with the jays, nutcrackers, and chipmunks for tidbits handed out by tourists.

Ravens seem to be playful birds, often soaring to great heights, sailing in circles above the mountains, while uttering their coarse croaks, doing half rolls and loops, falls, and then soaring again. For several years during my late summer visits to Grand Mesa in western Colorado to do research on pocket gophers I watched the ravens do aerial maneuvers a mile or two east of Land's End. The elevation here is approximately 10,000 feet. On clear days the ravens would start their croaking and aerial evolutions from this high place. Gradually they would climb in flocks of a dozen or more until they were mere specks in the sky. Their gyrations struck me as being the antics of a playful gang of birds rather than flights to find food such as the vultures perform.

In the desert mountains many lesser birds are seen and are

easily recognized. A few, not so well known, are rare visitors from Mexico. In the Chiricahua highlands and the Huachuca Mountains a variety of warblers appear. Among these are olive, pileolated, hermit, and grace warblers and races of Audubon warblers. The red-faced warbler is most likely to be seen in the overstory of high conifers. The Colima warbler, a bird that winters in Mexico, breeds in the Chisos Mountains in Texas. It is found nowhere else in the United States. In contrast with this rare visitor, the Mexican junco (*Junco phaenotus palliatus*) is likely to appear as an unwary visitor around your camp anywhere in the Engelmann spruce, Douglas fir, or aspen forests in the Arizona mountains. This Junco is the one with the orange-yellow iris. It also walks rather than hops.

When we consider all the different kinds of birds in the mountains, their behavior, their choices of environments, and why they happen to be where they are, we are confronted with a puzzle that involves ecological adaptation, genetic origin of races, and geological time. The first land vertebrates, the salamanders, crawled out of the water during the Devonion Period some 350 million years ago. The ferns and seed plants also began their differentiation and gave promise of floristic variety and varied habitats for animals that began to appear on the land. The first birds and mammals appeared in the Jurassic.

When flowering plants arrived in the Cretaceous, insects were there to pollinate them, birds were there to eat the insects, and new food chains were established. Then for millions of years, floods, drought, mountain formation, and repeated subsidence and elevation of the land was the story of the earth. Birds and animals either perished or lived, depending on how they adapted to the changing environments.

Birds that were genetically pliable took on characteristics that enabled them to live in new surroundings. Finally, when the great glaciation came during the Ice Age the northern birds moved southward with the coniferous forests in which they were accustomed to nest, or they found niches in deciduous forests at lower elevations. When the glaciers retreated, the birds moved northward to new forest combinations, including oak, beech, maple, hickory, birch, and pine woods. In these new habitats the surviving species were those with the biologic plasticity to change their behavior so as to find food, nesting places, and means of escape from enemies. Their very diversity contributed to the vitality of bird populations.

Conejos Peak in southern Colorado supports sedges, grasses, and dimunitive shrubs near the 13,000 foot summit. A packhorse trail leads right over the top of the mountain.

10

Never Summer Land

IF YOU GO HIGH ENOUGH in the Presidential Range in New Hampsire, in the Rockies, or in the Sierras of the far west, you will come to an elfinland of pigmy trees marking the high timberline of the mountains. Upward and beyond lies the Alpine Zone, treeless, windblown, dangerous in storms, serenely beautiful on quiet summer days, changeable by the hour, and entrancing because of the adaptations of plants and animals living in its demanding environment.

In the Southwest, in the San Francisco Mountains of Arizona, you will have to go high—11,000 feet or more—to reach the lower edge of the Alpine Zone. On Long's Peak in Colorado, timberline also occurs at about 11,000 feet. Treeline on Granite Peak northeast of Yellowstone National Park reaches approximately 10,500 feet. In the Canadian Rockies near Jasper, timberline is not much above 7,000 feet. In northern Alaska it comes down to sea level and all across the continent the northern timberline merges with the Arctic tundra.

The tension zone or transition area between timberline and alpine tundra fluctuates even in local areas due to wind exposure, snow depth, fog, and summer precipitation. In Glacier National Park and the Canadian Rockies timberline does not occur as an unbroken band of trees below the zone of shrubs, grasses, sedges, forbs, lichens and mosses. The erratic arrangement of mountain peaks, changeable winds, and temperature variations associated with discontinuous mountain slopes allow trees to grow at higher elevations in some places and the alpine tundra to reach lower in other places.

The transition zone between upright trees and alpine tundra is a fitting introduction to the world of flower strewn meadows, sedge bordered bogs, willow fields, glacial cirques, and rocky summits. The elements of subalpine vegetation have a close affinity with the alpine vegetation. In the Rockies this zone is commonly grassy with luxuriant stands of Thurber fescue, sheep fescue, or tufted hairgrass. Tall, spectacular forbs sometimes dominate the meadowlike aspect in mixture with patches of willows and other shrubs. Decumbent, scrubby, and dwarfed alpine firs, spruces, and whitebark pine, or limberpine in some places, appear in sheltered spots or grow in mats. Some of these trees are hundreds of years old. Although ice crystals blasted by the wind injure the twigs that grow above protecting rocks or out of the mat of prostrate branches, drifted snow protects the tree trunk and larger branches within the mat itself. In areas of less severe crystal scour, the trees are flagged and show branches and needles only on the leeward side.

Up and beyond this bizarre landscape lies the true Alpine Zone, a land of many contrasts. In spite of its severe physical and climatological conditions this never summer land is appealing and inspiring to many people. It attracts skiers in winter and mountain climbers, backpackers, photographers, and nature enthusiasts in summer. Some of the high altitude lakes provide excellent fishing and the many habitats provide homes for singular species of plants and animals.

The wide variety of sites includes ridges bare of snow in winter, rock fields, avalanche slides, cirques or glaciated ampitheatres, bold promontories, serrated peaks, and peaceful meadows. The profusion of land forms and dwarfed vegetation makes one forget that the world of forests, deserts, plains, and prairies lies far below and beyond remote horizons. The near landscape seems real and dynamic because of rock falls, snowslides, and heavy spring runoff, all of which are evidences of continuing geological erosion. Stinging windblown sleet and unpredictable snowstorms are common in every month of the year.

I remember one of these storms at nearly 13,000 feet on Conejos Peak in the Rio Grande National Forest in southern Colorado. There were five of us with saddle horses and pack mules. Suddenly the bright sunshine vanished. Within minutes we were enveloped in semidarkness while stinging sleet rasped our hands and faces. Then we were enveloped in St. Elmo's fire while lightning flashed and thunder rumbled on the slopes below the summit. A weird crackling or frying sound of static electricity was all around us. A bluish glow surrounded the head of my horse and his hair stood all on end. In this frightening scene we dismounted and lay flat on the ground to diminish the chance of being struck by lightning. But the storm soon passed and again we were in warm sunshine beneath a brilliant sky.

Recently I walked the nature trail with friends on Trail Ridge in Rocky Mountain National Park. It was one of those rare shirt-sleeve days in September. White billowing clouds drifted slowly in the azure sky. Hoary marmots wandered lazily among the sedges while pikas squeaked from their lookout positions in tumbled boulder patches. A raven drifted slowly over Forest Canyon, apparently surveying the view 2,000 feet below. We saw no ptarmigan, probably because of their perfect camouflage among the lichen-flecked granites and the brownish tints of plants beginning their senescence after a recent snowstorm. The Rydbergias had been frosted but their drooping sunflowerlike heads still retained traces of summer yellow. Still conspicuous were the Arctic gentians (*Gentiana algida*) with greenish flowers open to the translucent sky. In the clear light the mushroom rocks exhibited their strata of different ages in contrasting colors. I was reminded once again that the material of these rocks, at more than 12,000 feet, once was below the sea. It was a day to remember, for less than three weeks later, a Pacific storm front whitened Long's Peak and all the high mountain peaks of the Rockies.

These changeable weather and climatological conditions account for vegetation differences in the three main areas—northeastern, Rocky Mountain, and western—of alpine country in the United States.

The northeastern alpine segment occupies the highest summits in New England. Growth conditions for plants and animals are severe; winds of hurricane force blow in any month of the year; storms appear suddenly, and many are accompanied by snow or sleet.

On a recent October visit to Mount Washington in New Hampshire I saw rime, or frost formations, on posts and buildings built out two feet and more in a single night after a moderate snowstorm. Here, the highest wind velocity—231 miles per hour—ever recorded on a mountain occurred in April, 1934. Winds of one hundred miles per hour are common throughout the year. Timberline elevation here is markedly influenced by exposure to the winds. Plant species able to endure this tempestuous alpine climate are remarkably few as are mammal and bird. Most of the summer resident birds in these northeastern mountains nest in the forests below timberline where they largely es-

Mount Washington in New Hampshire has a severe climate. Storms may occur any day in the year. A light snow and strong winds built these rime shields on posts in early October.

cape the dangers of wind, fog, sleet, ice, and the chill factor of wind and low temperatures.

The Rocky Mountain segment of the alpine zone is less severe, climatically, even though it occupies extensive areas above 8,000 feet in the north and 11,800 feet in the south. Summer showers, sometimes daily, characterize this alpine zone of the high mountains from the Great Basin to the Great Plains, from the Canadian border to the Sangre de Cristo mountains in New Mexico, and to the San Francisco Peaks in Arizona. Winter snow is heavy but much of it accumulates in drifts on leeward slopes and in deep valleys. Temperature under the drifted snow drops only a few degrees below freezing and plants thus are protected from excessively low temperatures and winter winds. Plant species in the flora are numerous. And mammals, including elk, bighorn sheep, marmots, weasels, pikas, pocket gophers, and even white-tailed prairie dogs are present at elevations of 12,000 feet or more.

The western alpine segment occurs on high mountains from northwestern Montana to the Olympic Mountains on the Pacific Coast and southward through the mountains of Washington, Oregon, and California to its southern outpost on San Jacinto Mountain in California. The Sierran alpine flora has a paucity of tundra species in comparison with the Rocky Mountain alpine flora. Climate is the controlling factor. Summer rainfall is less frequent and abundant in the Sierras; soils are shallower; and barren rocky slopes and precipitous mountainsides are characteristic. Also, summer drought in the Sierras possibly favors the presence of annual species, whereas the Rocky Mountain climate allows the development of long-lived perennial plants which build the tundra mat and deeper soils.

There are relatively few places where one can reach the alpine zone by automobile. To experience all its variety and enchantment one must be an ardent hiker willing to endure the pressure of the wind, the sudden sleet storms, the terrifying strokes of lightning, and the pounding heart at high elevations. In my years of traveling in the mountains I have found that horseback travel with companions and a well equipped pack train is the best way to reach the remote alpine highlands miles from any road or town.

It is an easy ascent by automobile to alpine country on Mount

Washington in New Hampshire. If you wish to visit other peaks in the Presidential Range or Mt. Katahdin in Maine you will have to walk. Automobile roads lead to the tops of Pikes Peak and Mount Evans at more than 14,000 feet in Colorado. Trail Ridge Road in Rocky Mountain National Park permits exploration of alpine tundra over 12,000 feet. Tree limit is just above road level on Monarch Pass. Other high roads go over Independence Pass and Loveland Pass in Colorado. High elevation country with a prairielike appearance is accessible from the highway between Cooke City and Red Lodge, Montana. On Bear Tooth Pass, elevation 10,940 feet, low stature willows have not been excessively grazed by sheep. They add to the pleasing variety of other plants—sedges, grasses, and colorful forbs.

Alpine country in California may be seen by walking up the slopes above some of the roads through the mountain passes. On Carson and Sonora Passes, for example, you can see white bark pines distorted by the wind at timberline. If you are a confirmed hiker you can walk the Pacific Crest Trail through alpine landscapes that include rocky flats, talus slopes, serrated peaks, glacial basin meadows, and alpine lakes and ponds with water loving plants around their margins. Farther north, in Oregon and Washington, foot trails lead to alpine country on the slopes of majestic volcanic mountains including the Three Sisters, Mount Jefferson, Mount Hood, Mount Adams, Mount Saint Helens, and mighty Mount Rainier which towers to 14,410 feet. The trail up to the Hoh River in Olympic National Park to the glaciers on Mount Olympus provides a dramatic series of life zones, beginning among the mosses and ferns in the rain forest of gigantic Douglas firs, Sitka spruces, western red cedars, and Pacific silver firs, and leading to the glacial moraine above Blue Glacier. In this Olympic Mountain country you may also see rare or endemic plants which grow nowhere else in the world, including crazyweed (*Oxytropis olympica*), tower mustard (*Arabis olympica*), Piper bellflower (*Campanula piperi*), and the wallflower (*Erysimum arenicola*).

In all these far western alpine worlds, however, do not expect to see the endless variety of tundra species that grow on the millions of acres of Rocky Mountain alpine landscapes. These far western peaks stand more or less as isolated islands with limited space for alpine flora. Their rugged cliffs of granite or their loose structured volcanic soils do not permit development of extensive alpine tundra.

Alpine landscapes are marked by a wide variety of environments. The treeless terrain includes high peaks, bold promontories, windswept ridges, glacial cirques, sheltered coves, rock fields, grass and sedge-covered slopes, wet meadows, snow banks, and gin clear lakes. In many localities the alpine scene resembles a rolling prairie. But in spite of many alpine environments, four principal habitats, each with its characteristic plants, are easily recognized: rocky peaks, summits, and slopes; bog or marsh communities; willow fields; and alpine meadows.

Closer examination of each of these principal habitats generally reveals a fascinating, and sometimes perplexing, mixture of local surface features which offer selective microenvironments for plants, small mammals, birds, insects, and spiders. On a broad scale are the talus slopes composed of large rocks and stone debris accumulated at the bases of cliffs through frost action, temperature changes, and gravity. Scree consists of smaller rock fragments lying loosely on slopes. Lichens and mosses are the principal plants found on sheer rock walls and on the talus and scree slopes, although isolated plant clumps find a footing in protected crevices.

Fellfields, common in the Rockies and the Sierras, are areas covered with gravelly soil from which the tops of boulders protrude so that only part of the area offers substrata for plant growth. Fellfields sometimes grade into boulder fields where massive rocks are jumbled together. In these disordered rock piles little vegetation grows, but marmots, pikas, voles, and weasels find shelter there.

On gentle slopes, with gravelly soils or organic mulch, slips and slides produce natural terraces. Upright willows occasionally grow on the brows of these terraces and the faces are often covered with grasses, sedges, and forbs with showy flowers. Mass movement of soil down slope—a process called solifluction—happens when water from melting snowbeds renders the soil amorphous over smooth bedrock or over frozen substrata. Washing away of soil particles in rills that emanate from snow banks exposes stone streams and stone stripes, especially on steep slopes.

A boulder field in the alpine zone in the Rocky Mountains. Lichens on the rocks are the principal plants. Pikas, marmots, and weasels live in these rocks at the edge of alpine meadows.

Solifluction, frost heaving, and various other forms of water action are factors in the formation of rock circles and rock polygons. These features are so symmetrical they almost give the impression of being man-made. They are created by a sorting process whereby the finer materials are filtered out and removed from the centers of these rock patterns. When the polygons touch they form honeycombs or rock nets. The spaces between the polygons frequently are densely vegetated with bluegrasses, sedges, clovers, and shrubs such as skyland willow (*Salix petrophila*), snow willow (*S. nivalis*), or summit willow (*S. saximontana*). Dryas islands are low shrubby circular communities, several feet in diameter, formed by Mt. Washington dryad (*Dryas octopetala*). These patches of vegetation usually are surrounded by gravel or by small stones.

Alpine soils are usually stony or gravelly. But soils with high humus content develop where moisture is abundant and vegetation is undisturbed by big game animals or trampling and overgrazing by livestock. Disturbance of the soil layer results in slips and slides and accelerated formation of terraces which are a common feature of alpine landscapes.

Although many natural forces inherent in the alpine environment contribute to formation of local features such as rock streams, alpine deserts, sod peeling, soil terraces, and erosion gullies, some of these landscape aspects are manifestations of past overgrazing by domestic sheep. Damage to mountain top vegetation in the west was widespread, beginning as early as 1850 and reaching a peak by 1900. In seven years of study of alpine grazing indicators, Herbert E. Schwan and I found evidence of range deterioration in the high country from Montana to New Mexico.

Some of the damage resulted because ecological relationships were unknown and range management principles had not been developed. The policy of one supervisor of a National Forest in Colorado in the early days was to permit enough sheep in the alpine country to kill all the willows so more grass would grow. He succeeded in killing the willows. He also succeeded in scalping the slopes with sheep trails, breaking the sod, and trailing the snow banks with excessive numbers of sheep with consequent development of gullies and increased sedimentation of rills, creeks, and streams. Evidence of this kind constantly reminded Schwan and me that interpretation of alpine and other mountain floras must include consideration of man's influence as well as the natural causes of climate, weather, topography, snow, and the adaptations of plants that allow them to grow where they grow.

An awareness of plant adaptation adds to the pleasure of alpine excursions. The lure of colorful flowers is a satisfying experience in itself. But an understanding of the basic patterns of alpine tundra and how plants meet the demands of various habitats makes the seemingly complex mixture of plant communities ecologically meaningful and enjoyable to contemplate.

Alpine plants are adapted to withstand high winds, drought, extremes of temperature, and a short growing season. Consequently, species that have survived through the centuries are adapted for life in the micro-climate near the ground. Small size is almost a universal characteristic. On fellfields where the surface is rocky and little soil is present, cushion or mat plants are common. Some of these, such as the moss campion (*Silene acaulis*) have deep tap roots. These anchor them firmly against the wind and also supply moisture from deep in the substratum. Moisture is not always abundant in fellfields since dry winds blow in the summer and sweep these areas bare of snow in the winter.

Cushion and mat forms of growth are compact, retain heat from the sun, and hold soil in place. Evaporation of moisture from rain also is retarded. Typical cushion plants of the Rocky Mountains are the alpine sandwort (*Arenaria obtusiloba*), dwarf clover (*Trifolium nanum*), and Rocky Mountain nailwort (*Paronychia sessiliflora* var. *pulvinata*). The latter grows on barren rock surfaces where cracks allow penetration of the deep roots.

Many alpine plants of small stature have small or needlelike leaves which reduce evaporation. In contrast, however, the yellow stonecrop (*Sedum lanceolatum*) has succulent leaves. These retain moisture and enable the plant to endure dry periods as do the succulent cacti in the desert or the plants in our rock gardens at home.

The rosette form of growth makes some of the alpine plants quite as successful as the dandelions in our lawns. Rosettes expose maximum leaf surface to the sun and tend to avoid the dessicating effects of wind. Plants such as the snowball saxifrage (*Saxifraga rhomboidea*), commonly found in alpine meadows, produces a basal rosette of leaves where the air near the ground is usually warmer than the air above the plant. The tundra dandelion (*Taraxacum ceratophorum*), along with our old friend the common dandelion (*T. officinale*), occurs in valleys, meadows, and in the alpine zone. The bog-rooted spring beauty (*Claytonia megarhiza*) is a beautiful rosette plant with fleshy leaves and light pink or white flowers. It produces a large tap root that grows to a length of five or six feet.

Hairy coverings also are an adaptation of some alpine plants. These coverings vary with the species and may be woolly, spiny, short, long, tangled, or sparse. Rydbergia (*Hymenoxys grandiflora*), a showy alpine sunflower that dies after blooming, for example, is densely covered with woollike hair. In Colorado it occurs between 9,000 and 14,000 feet. Translucent hairs grow on the catkins of alpine willows. These hairs allow the sun's rays to penetrate while they retain heat and decrease transpiration.

The "snow loving" plants in the alpine zone always fascinate the casual visitor because of their apparent adaptation to cold. Some of these, such as the snow buttercup (*Ranunculus adoneus*) and the snowlily (*Erythronium grandiflorum*), burst through the snow or come into flower as the snow melts and recedes up the slope. Actually, these plants are less exposed to extremely low temperatures than are plants in the fellfields and open meadows where the snow melts earlier in the season. Temperature under more than two and one half feet of snow almost never drops below 27° F. The physiological fires within the bulbs, roots and stems raise the temperature of the plants allowing them to grow beneath the snow. Thus they are ready to burst into bloom when they are finally exposed to the warmth of the sun.

Grasses, sedges, and rushes are abundant in the alpine zone. These, like practically all other tundra flowering plants, are perennial. The short growing season and the severe climate at high altitudes preclude the possibility of regular annual seed crops. Perennials, on the other hand, can grow and develop over periods of many years and then come to fruition in favorable seasons. Many of the perennial grasses and sedges also spread by vegetative reproduction with runners like those of strawberries or with rhizomes which are underground stems.

Although grasses and sedges are especially abundant on moist alpine slopes and in meadows the casual observer is inclined to give them little notice since they lack showy flowers. The rushes are grasslike in appearance but they have flowers with six-parted scalelike perianths (calyx and corolla together) which make them close relatives of the lilies. At high altitudes they are low in stature, sometimes growing only a few inches to a foot or so in height. Rocky Mountain species are Drummond rush (*Juncus drummondi*) and Parry rush (*J. parryi*). In New England the highland rush (*J. trifidus*) grows in dense tufts six or eight inches high. It can be seen along the road near the summit of Mount Washington in New Hampshire.

If you wish to be more than a casual observer of the alpine flora you will find reward in a hand-and-knee expedition among the diminutive plants. Not only will you avoid some of the wind, but a close-up examination of even a square yard of area will reveal the surprising abundance and variety of species keeping company with one another.

Identification may not be easy since many of the alpine plants grow only above timberline and are not often seen by outdoor visitors. Some high altitude plants, however, are adapted both to the mountain tops and to the sub-alpine timber zones below. Others, such as the yarrows and dandelions grow all the way up and down the mountains. And many, including the willows, azaleas, sedges, grasses, and goldenrods, have plant relatives that grow near and far, from lowlands to mountain tops. The resemblance between these familiar species and their dwarf alpine counterparts can give you at least a feeling of familiarity with some of the mountaintop species.

A general acquaintance can be made with the alpine and sub-alpine plants through use of paperback guides and pamphlets available at nature museums, local book stores, and National Park information headquarters. Some of these guides, such as *Alpine Zone of the Presidential Range*, by L. C. Bliss, describe alpine plant communities and give aids for identification of some of the plants above timberline. If you drive above treeline in the Colorado Rockies, *Alpine Flowers of Rocky Mountain National Park* by Bettie E. Willard and Chester O. Harris, illustrates in color some of the common wildflowers according to the major ecological types in which they occur. In *101 Wildflowers of Glacier National Park,* by Grant W. Sharpe, the flowers or ripe fruits are arranged by their colors, including blue, green, pink, red, white, yellow and black. The plants described and illustrated in this booklet include both alpine species and those that grow in the forested zones at lower elevations.

Flower picking without written permit is prohibited in national parks, many recreation sites, state reservations, national forests and other public areas. In general, these restrictions are reasonable since many rare species would vanish if unrestricted collections were allowed. I was amused a few years ago, however, when I heard a forest ranger reprimand a woman for picking a few columbine flowers in Utah while a nearby band of 1,000 sheep was mowing them down by the millions. But restrictions are necessary, especially in areas where thousands of recreationists come to see and enjoy natural beauty. This is particularly true in the alpine country where the plants are fragile and can be killed even by moderate trampling. Once killed, some of the alpine species may not be replaced by natural succession in fifty to one hundred years.

If you want the whole picture of the alpine flora you should observe the lichens and mosses. These are perennial plants, some of which are adapted to the most severe environmental conditions imaginable. Some species grow on soil; others grow on rocks where they endure extremely high and low temperatures, limited moisture supply, and blasting by snow and sand. Many are showy and add elegance to what otherwise might be a colorless landscape.

Most of the mosses require a hand lens for proper identification. The haircap moss (*Polytrichum juniperinum* var. *alpestre*), however, is easy to recognize even though it is a dwarfed variety of the genus which is widespread in the northern states. I used to collect the large species on the moist rocky hillsides in Wisconsin for my students in moss identification at Marquette University. The plants could be studied with the naked eye since the setae or stalks grew to a height of four to six inches and the hairy caps (calyptras) covering the capsules were conspicuous. When the caps were pulled off, the operculum, a circular lid on the capsule, was exposed. When the operculum was removed, the peristome, a fringe of fine teeth could be seen and the spores could be shaken out like salt from a salt shaker. Thus a part of the life cycle of a typical moss was easily demonstrated.

The lichens also are a numerous group and are abundant in some alpine areas of the Northeast and in the Rocky Mountains and Sierras. Lichens give rise to some of the attractive colors on rocks, not only in the mountains but even in deserts. These distinctive plants are a mixture of fungus strands and usually single celled algae. The latter possess chlorophyll which enables the algae to provide food for themselves and for the parasitic fungi which furnish moisture, inorganic nutrients, and shelter for the whole lichen structure.

Some lichens, especially those on rocks in exposed environments, have the appearance of paint stains. These are the crustose lichens. In more favorable habitats, lichens grow as foliose types, with structures resembling small leaves, or as fructose forms growing in mats with upright branches, forked stalks, and fingerlike appendages. Among the conspicuous lichens found at

Lichens are the pioneer plants on rocks from mountain foothills to the highest peaks. These specialized plants, consisting of algae and fungi, grow slowly and persist for many years. The blues, reds, yellows, and greens of lichens add much beauty to rocky landscapes.

high altitudes are the map lichen (*Rhizocarpon geographicum*) which forms a paintlike crust on rocks, the ring lichen (*Parmelia centrifuga*) which produces circular patches with dead centers, and the snow lichen (*Cetraria nivalis*) which produces tufts of stalks with broad lobes finely divided at the tips. Common at lower altitudes is the reindeer lichen (*Cladonia rangiferina*) with which many outdoor people are familiar.

The Northeastern alpine country is limited in extent. The most extensive area occurs on the Presidential Range of the White Mountains in New Hampshire. Among the highest peaks are Mount Washington (6,288 ft.), Mount Adams (5,798 ft.), Mount Jefferson (5,715 ft.), and Mount Clay (5,532 ft.). Timberline in this range fluctuates between approximately 4,800 and 5,200 feet, depending mostly on exposure. In sheltered spots behind rocks fir and spruce may exist a few hundred feet higher. Above timberline, the alpine zone is characterized by lichens, mosses, grasses, sedges, and dwarf shrubs. True alpine country also occurs on Mount Katahdin in Maine.

In the Adirondacks some forty-six peaks reach 4,000 feet or

more in elevation. The tops of most of these are forested. The original wooded summits of a few have been burned over and their more or less barren tops should not be considered to be alpine country. Alpine environments, however, do occur on the following peaks: Mount Marcy, Algonquin, Haystack, and Skylight.

The Northeastern alpine plants are not as numerous as are those in the Rocky Mountains but all show the dwarfing effects of high altitude climates. Many are conspicuous because of their large showy flowers. Diapensia, or Mountain bride (*Diapensia lapponica*) grows in dense leafy cushions and at flowering time produces a crown of pure white flowers on two-inch stems. It withstands the high winds on Mt. Washington and is found on the slopes of Algonquin in the Adirondacks.

A tundra plant that grows across the northern continent and on the high peaks in New England is Lapland rosebay (*Rhododendron lapponicum*). This dwarf heath, related to the magnificent rhododendrons found in the southern Appalachians and along the Pacific coast, produces cerise or royal purple flowers in conspicuous terminal clusters. The rusty leaves are only an inch long. In similar dwarfed fashion the Arctic azalea (*Loiseleura procumbens*) hugs the earth and raises its pink or rose-colored flowers scarcely an inch above ground. It occurs on Mount Washington but not on the Adirondack peaks.

Other prostrate shrubs add variety to the alpine flora of the Adirondacks and the far northern peaks of the Appalachians. The bearberry or arctic willow (*Salix uva-ursi*) is vinelike as is the crowberry (*Empetrum nigrum*) which produces runners with short needlelike leaves. The scales of the arctic willow catkins are rose-tipped and are covered with silky hairs. Its cousin, the dwarf willow (*S. herbacea*), circumpolar in distribution, is one of the smallest willows, seldom reaching an inch in height. Its flowers bloom in June soon after the snow melts. In contrast with these matted plants the mountain sandwort (*Arenaria groenlandica*) grows in tufts. Its white flowers can be seen along the coast of Maine as well as on the mountain summits.

Grasses and grasslike plants also form an important part of the northeastern alpine flora even though they lack brightly colored flowers. Alpine holygrass (*Hierochloe alpina*) is locally common on the Adirondack high peaks and also occurs in the Presidential Range. Deer's hair (*Scirpus caespitosus*) is a widely distributed bulrush growing in dense tufts up to twelve inches in height. It is conspicuous on the mountain landscape when it turns to straw colored clumps in autumn. A smaller but common plant is the highland rush (*Juncus trifidus*), easily distinguished by the three leaves extending outward from the tiny green flower at the tip of the stem.

An exhaustive search of the northeastern alpine country will, of course, reveal a plethora of other interesting plants, of which many grow in the timbered zones below treeline. Among these is the alpine bistort (*Polygonum viviparum*), a relative of the "smartweeds" we find along streams in the Midwest and sometimes among the vegetables in our gardens. The alpine bistort is viviparous, bears live young in a sense, in that the little bulbils produced on the stem begin growth before they fall to the ground. This enables the plant to reproduce in the same summer in which flowers and seeds are formed.

Anywhere above tree limit in the Rocky Mountains you may expect to see a kaleidoscopic mixture of alpine plants. Some of these are abundant and widely distributed. Others are restricted to specific habitats, such as protected spots beneath overhanging rocks, stream sides, snowbanks, or shallow ponds. One of these, the alpine quillwort (*Isoetes muricata*), is so inconspicuous it is frequently overlooked by plant collectors. It grows in bogs or lakes where its onionlike leaves are partially submerged. It intrigues botanists because of its phylogenetic position between the ferns and the clubmosses. It looks like a seed plant, but instead of seeds it produces sporangia in the hollow leaf bases with megaspores and microscores. It is widely distributed throughout the western states. I have found it on the Rio Grande National Forest at approximately 11,500 feet.

In contrast, dwarf willows, sedges, grasses, and showy flowering plants with broad environmental adaptations can be seen almost anywhere above timberline. American bistort (*Polygonum bistortoides*) with its white densely flowered terminal racemes grows in moist areas from 6,000 to 12,000 feet. Alpine avens (*Geum rossii*), closely related to the strawberries, but with erect yellow flowers, is one of the commonest plants in alpine meadows. Probably the showiest of them all is Rydbergia (*Hyme-*

The yellowbelly marmot sitting on his sentinel rock in the western mountains gives loud chirps to warn of danger. He emerges from hibernation in March before the snow is gone. Look for him in the rocks in western mountains up to elevations of 12,500 feet.

noxys grandiflora), with large solitary yellow flowers with densely woolly bracts. But the favorite of flower lovers is the alpine forget-me-not (*Eritrichium aretioides*). Its miniature mat of flowers seems to absorb and intensify the blue of the sky above.

An excursion among the snow fields in late July will reward you with the spectacle of plants in all stages of growth from seedlings near the snow to plants in full bloom a few feet away in the muddy fringes of giant drifts. The snow buttercups (*Ranunculus adoneus*) sometimes bloom before the drifts melt and uncover the leaves. The glacier lily (*Erythronium grandiflorum*), which resembles the yellow adder's tongue (*E. americanum*) that most of us have seen in rich woods east of the Mississippi River, produces whole fields of drooping yellow flowers following the recession of snow up the mountainsides.

Also widely distributed is the black-headed daisy (*Erigeron melanocephalus*), related to the asters, which grows in the spruce-fir zone as well as on snow accumulation areas in the alpine. The bracts of the flower head, beneath the white petals, are packed with woolly purple-black hairs. Parry clover (*Tri-*

folium parryi), one of the three common alpine clovers, forms carpets of leaves and reddish-purple or rose colored flowers in places where the snow melts early in the summer. Its relative, Alpine clover (T. dasyphyllum), forms mats in drier places such as fellfields and on gravelly soil. The globose flower heads with ten to twenty purple to pink or sometimes bicolored flowers are borne on long peduncles that raise the flowers above the leaves. In contrast, dwarf clover (T. nanum), common on fellfields grows in low compact mats with rose-purple heads consisting of two to three flowers borne just above the leaves.

In the alpine meadows scores of flowers bloom early; some are in bloom all summer; and a few are conspicuous in early September, even if frost has bronzed most of the vegetation. Among the early ones are the alpine primroses (Primula angustifolia) with leaves in basal rosettes and tubular flowers with purple, pink, or occasionally white petals. The mountain hairbell (Campanula rotundifolia) is hardly to be mistaken because of its blue nodding bell-like flowers. More conspicuous is the yellow paintbrush (Castilleja occidentalis) which grows in showy clumps at 10,000 to 13,000 feet. Even more striking is the kings crown (Sedum rosea) with its dense clusters of dark red or rose-purple flowers on stems up to a foot or more in height. In autumn the whole plant turns a brilliant red. It also grows in moist places in the Sierras, the Cascades, the Olympic Mountains, and northward to Alaska. Less showy, but attractive, is the alpine wallflower (Erysimum nivale) which resembles some of its yellow flowered relatives in the mustard family that grow so abundantly in abandoned fields at low elevations.

If you examine the alpine vegetation carefully, particularly on moist slopes and in meadows, you will see that the great volume of vegetation is produced by grasses and sedges. The two groups may be distinguished by the arrangement of leaves on the stems. In grasses the leaves usually alternate in opposite directions from the hollow round stems. In sedges the leaves grow in three directions from the solid three-sided or triangular shaped stems which have no nodes or joints. The flowers of grasses are rarely unisexual; instead, stamens and pistils are included in the same flower. In sedges the male and female flowers are often borne in different inflorescences or in different parts of the same inflorescence.

The fruit of sedges is formed in a perigynium, a saclike structure, often beaked with an opening at the tip. In the genus *Kobresia*, one of the important sod-forming species in the alpine climax vegetation, the perigynium is split down one side to the base. In the genus *Carex*, the perigynium is closed except for the apical orifice which may be toothed or split part way down.

More than 600 species of sedges grow in North America. In the Rocky Mountain region 165 species and many varieties occur. Some of these are adapted to high altitude environments where they are instrumental in erosion control, development of sod, and formation of alpine soils. Many are comparable to grasses in forage value for wildlife and domestic livestock.

Some of the sedges are widespread and grow over a broad range in elevation. Hayden's sedge (*Carex haydeniana*), for example, grows in mountain habitats from Alberta, Canada to Oregon and from Colorado to California at elevations varying from 6,500 feet to 14,000 feet. It is a hardy species with culms or stems four to twenty inches high, depending on favorableness of the habitat. It is common on alpine slopes. Two sedges found on sunny moist alpine slopes and damp meadows are native sedge (*C. vernuncula*) and black alpine sedge (*C. nigricans*) which remains green throughout the summer season.

New sedge (*C. nova*), handsome sedge (*C. bella*), and sheep sedge (*C. illota*) inhabit alpine meadows, the shores of alpine lakes and streams, and moist woods near timberline. On rocky slopes, dry mountainsides, and summits to 13,000 feet you may expect to find black-and-white sedge (*C. albonigra*), Arapahoe sedge (*C. arapahoensis*), and false bulrush sedge (*C. pseudoscirpoidea*). The latter produces stout woody rootstalks from which the culms arise; its spikes, or simple inflorescences, are solitary on the culm, erect, and linear—twelve to forty mm. long.

Kobresia (*Kobresia myosuroides*), the specialized one with the perigynium split on the side, is the soil producer in the alpine climax. It is grazed by elk, mountain sheep, and usually overgrazed by livestock. In pristine condition it forms solid sods or conspicuous communities which are continuous mats of green turf that turn to bronze, copper, or yellow gold in early autumn. One of the best stands of Kobresia I know is on Green Horn Mountain southwest of Pueblo, Colorado. This is horseback country at 12,000 feet where the tourists never go.

Men who settled in the mountains were true pioneers who loved the rugged life, built their own cabins, and were largely self-sufficient. They used the products of the land—wood, fish and game animals. Some were miners and many became successful raisers of livestock.

11

Men and Mountains

THE EARLY WHITE EXPLORERS in America encountered many Indians but few of them inhabited the mountains. Instead, they preferred the more equable climate of the coastal areas, or the shores of lakes and rivers where water, wood, and game animals were abundant. Even the inhabitants of the deserts were probably the descendants of ancestors who settled there when the climate was more humid and prehistoric animals were easily hunted.

Apparently the Indians were afraid of many of the foreboding mountain peaks. According to the legends, kings, fierce warriors, beautiful maidens, and even the thunder and lightning resided in the mountains. The great fire mountains of the West spread smoke, ashes, and heat over the land and scorched the skins of the ancient ones, while out of the clouds the Great Spirit called the braves to the Happy Hunting Grounds.

In spite of the supernatural powers abiding on mountain tops some of the Indians lived in or at the edge of the mountains. The Cherokee were numerous in the southern Appalachians in the eighteenth century, but even their language was derived from the Iroquois who were part of the Algonquin family of tribes of the north. The Western Indians avoided the steaming rumbling geyser area of Yellowstone as a place possessed by demons. Mount Rainier in Washington, discovered by Captain George Vancouver in May, 1792, had many Indian tribes around its base—Yakimas, Taidnapams, Nisqually, and Puyallup Indians. In 1833, the Indians guided Dr. William Fraser Tolmie, of the Hudson Bay Company, on a botanical expedition to Mount Rainier. Historical evidence, however, indicates that the Indians never climbed the mountain since it was a place of mystery not to be disturbed by human presence.

The Indians did use the mountain passes as routes of travel to hunting grounds. In New England, when white men began their settlements on the coast and in the valleys, the Indians used the gaps as war trails, even though legends and superstition warned them not to make their permanent abode in the White Mountains of New Hampshire or on Mount Katahdin in Maine. Much Indian lore is found in the basins of the Green Mountains of Vermont and in the White Mountains and the red men undoubtedly used Crawford Notch, Kinsman Notch, and others as passageways through the rugged terrain.

In the southern Appalachians Indians and many white explorers used the gaps in crossing the mountains. Some historians believe that Hernando De Soto and his Spanish conquistadores explored the Blue Ridge and Unaka country in 1539 and 1540. They probably used Wallace Gap in crossing the Nantahala Mountain range and Ooltewah Gap in crossing White Oak Mountain on the way to the Tennessee River near present day Chattanooga. In this same region Daniel Boone crossed various mountain ranges on Indian trails and explored passes, particularly Cumberland Gap.

The early explorers in the West avoided the mountains whenever possible since the goal in mind was gold, scouting for enemies, searching for low-level passageways to the Pacific Ocean, or determining boundaries of the Territories for the new developing nation. The Spanish largely avoided the Rockies in the eighteenth century. But in 1761 and 1765, Juan de Rivera led expeditions into the San Juan Mountains, the Green River plateau country of Utah and the Wasatch Mountains. Later, the Spanish adventurer Juan Bautista Anza, founder of Monterey and San Francisco, assigned Fray Silvestre Velez de Escalante the task of finding a supply route from Santa Fe to California. The party entered new and unexplored land in southwestern Colorado, traversed the Uinta Mountains to the Green River,

The Indians drew pictures on rocks depicting the kinds of animals they knew in former times. Some of these pictographs are believed to represent the location of animal trails and good hunting spots.

and struggled through the Wasatch to Utah Lake and finally back to Santa Fe.

The expedition that really opened the West to exploration and conquest was the journey of Lewis and Clark. The Rockies had been crossed previously in the far north by Alexander Mackenzie in 1793. But the journey of Lewis and Clark, 1804-1806, was the one that produced new information about the interior of the continent, planted the idea of claiming the Oregon country, and stirred the imagination of trappers, traders, hunters, and settlers.

Historic Naches Pass in the Cascades east of Mount Rainier was crossed from west to east in 1841 by Lt. Robert E. Johnson who was sent by Commander Charles Wilkes of the Wilkes Expedition. Then in 1853, James Longmire led the first wagon train across the Cascades north of the Columbia River. This led to the beginning of settlement near the present site of Tacoma, Washington, and the staking of claims along Puget Sound.

Explorers after Lewis and Clark entered the mountains, still with the idea of finding easy passage to the sea or settling disputed boundaries such as that of the Louisiana Purchase being contested by Spain. Zebulon M. Pike, in 1806, depending on faulty maps, followed the Arkansas River to the present site of Pueblo, Colorado, and then attempted without success to climb the Peak that now bears his name. He did travel to the mouth of the Royal Gorge, explore South Park, and cross the Sangre de Cristo range. But his report, issued in 1810, was inconclusive and raised more questions than it answered. Stephen H. Long, in 1820, followed the Platte River to the site of Denver and south to Pikes Peak. Long then went southward to the Arkansas River, encountering heat, storms, and Commanche Indians, and found nothing in the mountains to justify crossing the plains to reach them. To Pike and Long the plains were "The Great American Desert."

The great government expeditions of the 1840's culminated in the explorations of Capt. John Charles Fremont. These journeys were primarily to learn the nature of the rivers between the Missouri and the Rocky Mountains, to aid immigration through South Pass which was the crossing in Wyoming to Ogden and the Northwest, and to map the Oregon Trail. His expedition did publicize the nature of the western empire and his exploration of the Great Basin influenced Brigham Young who led the Mormons into Utah.

Before the great expeditions were completed, mountain men, including Kit Carson, Thomas Fitzpatrick, and trappers for the Hudson Bay Company knew the mountains from one end to the other. John Colter and Jim Bridger were fur traders who told tall tales of geysers, spouting springs, smoking valleys, and glass mountains. Their accounts seemed so improbable they were taken as gross exaggerations.

But the strange reports were true and the phenomena the trappers had described were seen in the early 1860's by Montana gold seekers and were examined by the Hayden Expedition which entered the Yellowstone country in June, 1871. The pioneer photographer William H. Jackson was with the party and his incomparable photographs caught much of the pristine beauty of the landscape that later was to be established as Yellowstone National Park. John Merle Coulter, under whom I studied at the University of Chicago, was the young botanist of the expedition. He told me how he was assigned to watch one of the least promising of the geysers. During the night it erupted and showered his sleeping bag with hot water. It turned out to be one of the largest geysers in the basin!

The fur trade in the Rocky Mountains had diminished before the explorations of Yellowstone and other western mountain regions were made. All the trapping, however, was not done in the northern mountains. The fur trade also flourished in the Southwest. In 1821, with the independence of Mexico, wagon trains traveled the Santa Fe trail and the village of Taos boomed as a fur trade center that rivaled those of Vancouver and St. Louis. Among the trappers who operated out of Taos and Santa Fe were Kit Carson, Pegleg Smith, James Ohio Pattie, "Old Bill" Williams, and the four Robidoux brothers. These mountain men led trapping expeditions into Colorado, Utah, Arizona, and California.

The trappers of later years must have worked hard for their money. Some of the prices quoted by J. and A. Boskowitz, 246 Lake Street, Chicago, in 1877, were as follows: mink, 5–75¢; muskrat, 3–10¢; beaver, 50¢ to $1.00; wolverine, $3,00; deerskin 70¢; elk, 5–15¢ (Indian dressed); bear, $2.00; grizzly bear, $5.00; and buffalo robes, $2.00–$5.00.

The real conquest of the mountains began with the discovery of gold. The Spaniards, with dreams of golden empires to pillage and destroy, failed because there was little or no gold in the lands they explored. Vasquez de Ayllon, in 1526, attempted to found a settlement in the southeastern United States while looking for gold and jewels. Apparently he died in the wilderness of South Carolina. Captain Pánfilo de Narváez in 1528 brought more colonists when he landed at Tampa Bay. They found a few gold trinkets among the possessions of the now extinct Apalachee tribe who apparently obtained them by barter with Indians in Georgia where nuggets were present in the river sands. But Narváez also came to an untimely end.

Then Cabeza de Vaca brought tales of the Seven Cities of Cibola. This led to the fabulous journey by Francisco Vasquez de Coronado and disillusionment, for there was no gold in the Grand Canyon nor on the plains of Kansas. Far to the west,

Juan Rodriguez Cabrillo, in 1542, sailed along the California and Oregon coast, died in 1543, and was succeeded by his pilot, Bartolome Ferrer, who noted the high mountains that lifted their snow covered heights above the sea. They found no gold, such as the earlier Spaniards had looted in the Inca empires, nor did they realize that behind the Coast Ranges lay gold deposits that would change the course of history in a nation not yet founded. Even Lewis and Clark never realized they were in gold-bearing territory when their expedition moved through Montana.

Then on January 24, 1848, James Marshall found flecks of gold in the tailrace of John Sutter's sawmill on the American River in California. The stillness of the mountains was broken. In 1849, from all corners of the earth, men came in wagons, on foot, on horseback, across the Isthmus at Panama, and by boat around the Horn, until the forty-niners numbered forty thousand in the foothills with additional thousands swarming over other parts of what would become California.

The drama unfolded with the coming of prospectors, builders, capitalists, whiskey, honky-tonk girls, new towns, new roads through the passes, and ultimately the transcontinental railroad. The country itself was relatively undisturbed at first as miners panned for gold. Then mining evolved into all the destructive practices gold mad people could devise.

Water was diverted to serve flumes and sluices in one gulch after another. Hydraulic mining then began to wash ore from its beds with powerful jets of water. Placer mining, washing the surface soil to leave the heavier metals, devastated the landscape and filled the streams with silt. Stamp mills shook the earth day and night crushing the ore so gold could be extracted. Then giant dredges, floating like landlocked ships, began piling giant ridges of tailings along the Yuba River in 1850 and did not stop until 1968 when Yuba Consolidated Industries and other companies ceased operations. With depletion of the surface gold the industrial effort then changed to deep mining with shafts and tunnels that crept hundreds of feet into the earth.

From 1850 until 1900 additional discoveries of gold and silver were made from Arizona, Nevada, and Colorado to South Dakota, Oregon, and Alaska. In 1851 discoveries were made in the Rogue River Valley in Oregon. The rush to the Fraser River in British Columbia was made in 1858. Then, in 1859 the "Pikes

The Silver Lake Mill in the San Juan Mountains fell into disrepair many years ago. If the price of gold remains high there could be a return of the mining industry. The mountains still hold fabulous quantities of minerals.

Peak" rush to the Rockies began. Central City became one of the major boom towns in Colorado and it attracted many notable people, including President Ulysses S. Grant, Mark Twain, and Horace Greeley. Stage performances began in 1861 and the famed Teller House, revived in the 1930s, now opens each summer with theatrical performances attended by people from throughout the nation.

Silver also started some of the great bonanzas in the West, including strikes at the Comstock Lode and Virginia City, Nevada, Globe, Arizona, Silver City, New Mexico, and in Colorado's San Juan Mountains. Dozens of towns were built in this

"Switzerland of America," many of which became ghost towns in later years. Towns that still survive in the San Juan country include Silverton, Durango, Telluride, Ouray, Ridgway, and Lake City. Hundreds of the early towns boomed, attracted their unruly characters, shady ladies, saloon keepers, and hard rock miners. Then when the gold or silver ran out, the houses and mines were abandoned and only the ghosts of their inhabitants remained. The names of the mines are reminders of the hopes and aspirations of the miners and the spirit of the times—Holy Moses, Tomboy, Gold Cup, Yankee Girl, Molly Gibson, Camp Bird, and Blistered Horn.

A perusal of some of the Ghost Town guide books, many of which contain photographs taken in the 1870s, 1880s, and 1890s, will show how men tinkered with the mountains. They burrowed into the rocks, left slag piles, washed away the soil with their placer jets, and denuded the forests for mine props, railroad ties, houses, mine buildings, and fuel. In the old photographs the mountainsides are barren of trees. Repeat photographs, taken in the 1960s, show regrowth of pine, spruce, and fir forests and aspen successions where fire burned the mountainsides. The ecological damage from early mining now is diminishing as the ghost towns crumble into dust. Those that remain are reminders that men's struggle for riches in the mountain wilderness led to the development of the modern West and the American nation.

Now vandals are dismantling the ghost towns piece by piece and carting them away in campers, jeeps, and trucks. One reason is the soaring market value of antique bottles, mustache cups, lanterns, miner's tools, wagon wheels, and opium scales. Even the weathered boards from houses and gold mills are selling for a dollar per board foot to arts and crafts workers who decorate them with cones, weeds, shells, or rocks. One team of pot hunters flew into the Inyo Mountains in California with a helicopter and practically dismantled an entire town. Theft laws of the states and the federal Antiquities Act of 1906 seem to have little deterrent effect. Most people, however, are still content to walk around and photograph the old towns, and possibly reminisce about a kind of man's use of the earth that is rapidly becoming a legend.

It has been said that the Winchester and the .45 Colt conquered the West. But it was the railroads that conquered the mountains. The rails brought fortune, frustration, growth, depression, booms, and busts to the mountains. They opened new areas for industry, created towns, changed the prices of commodities, and moved people into an environment unlike that known to the rest of the nation.

Railroads boomed the mining industry, introduced agricultural practices into the high hills, and made possible the denudation of the primeval forests. Successful logging in the Rockies, the Sierras, and the Northwest was made possible by the power of the iron horse. The height of railroad logging was reached about 1900 and lasted until the 1930's. Then log trucks with

Ghost town near Independence Pass in Colorado. Many of these towns were abandoned when the mines played out.

greater mobility took on the task of moving the woods to the sawmill and the lumber products to the markets.

The obsession for railroad building reached its peak in Colorado with its multitude of high mountains. The building of the railroads involved wild dreams, great expenditures and expectations of wealth, defeat, and ultimate success over almost hopeless odds. Construction workers battled blizzards, thin air, homesickness, ground slippage, rock cave-ins, financial privation, dynamite explosions, suffocation from gas and smoke in tunnels, burned snow sheds, and the wrath of wives directed at their wayward men when girls and whiskey were brought in by saloon keepers.

Some of the transcontinental railroads avoided the mountains. The Union Pacific crossed the plains of Wyoming and followed the valleys and low passes. The Denver Northwestern and Pacific Railway, the "Moffat Road," named after David H. Moffat, took the hard way—Denver to Gilpin, Steamboat Springs, Craig, Colorado, and Jensen, Heber, Provo, to Salt Lake City, Utah. There were twenty-seven tunnels drilled in hard granite in twenty-four miles, with many trestles over gulches between tunnels. The rails reached 11,660 feet over Rollins Pass. When the engines were not buried in snow or derailed with ice in winter they pulled cars loaded with sheep, wool, cattle, coal, and gilsonite ore from Craig.

Before the turn of the century the Denver and Rio Grande Western had built 1,300 miles of narrow gauge track in Colorado. The locomotives burned coal and wood. On November 10, 1872, the Rocky Mountain News noted that buffalo chips also were used for fuel. The products hauled along the east front of the Rockies included limestone from the Monarch quarry for manufacture of steel at Pueblo, crude oil from Chama, New Mexico, lumber, machinery, and people. The locomotives weighed 25,000 pounds at first; later they weighed as much as 93 tons. The ride was rough, the time of arrival was unpredictable, and even the passengers required fortitude and a strong constitution. Only a few years ago, on the narrow gauge road from Alamosa to Durango, one of my friends asked the conductor when the train would get to Durango. The conductor's reply was, "What do you think I am, a goddamned prophet?"

One of the fabulous jobs of railroad bulding was the project of Otto Mears to build a railroad between Silverton and Ouray

The narrow-gauge railroads have practically disappeared, and with them has gone the romance of the steam locomotive with its deep throated whistle that echoed up and down the canyons. This one used to run through the gorge east of the Black Canyon of the Gunnison River in Colorado. Now the Blue Mesa Reservoir covers much of the valley between Cimarron and Gunnison.

in the majectic San Juan Mountains. He built twenty-eight miles of track and came within eight miles of his goal when the impassable Uncompahgre Canyon stopped the venture. Undaunted, he backtracked through Durango, Rico, and Telluride to join Silverton to Ouray. The job required 217 miles of track.

The switch backs, interminable slopes, and the deep canyons made railroading in the mountains a preposterous and exciting business. On the long down grades, before air brakes were introduced, an indispensable piece of equipment for every brakeman was his pick handle or brake club. This piece of stout oak or hickory was inserted between the cross bars of the brake wheel on top of every freight car. The club added leverage so the brake staff could be twisted to wind up the brake chain which pulled the lever that pushed the brake shoes against the car wheels. If the brakeman failed at this job the train was due for a wild run down the mountainside.

Some of these downhill and uphill grades with endless curves and switchbacks made the going slow. At Gold Hill, west of Boulder, Colorado, you could stand at the depot and see the train appear at the same level across the canyon and hear its whistle. Then it would begin its slow descent into the valley from which it would finally climb again to your side of the mountain. That would be four hours after you first heard its whistle.

The logging railroads of the western mountains represented another achievement in development of specialized transportation. The stepped-up demand for timber in the 1880s and in later years introduced the donkey engine, a small stationary steam engine, for unloading logs at the mill pond. Later, the engine took over the task of skidding logs out of the woods—a job formerly done by horses or oxen and gravity. With this increased capacity for handling logs and with the mainline railroads available for long hauls from the junction points, the coming of the logging locomotive and narrow gauge tracks to negotiate sharp curves became inevitable.

The first locomotives were gypsy-rigged affairs with inadequate power to pull cars loaded with logs that sometimes weighed more than the locomotive. One innovation, made by John Dolbeer of Eureka who patented the donkey engine in 1882, was attachment of the donkey engine to the pilot, or cowcatcher, of the locomotive. Thus a portable supply of steam from the locomotive was available for yarding logs from the forest to the rails.

Gradually the size and power of the locomotives increased and many famous ones received names as well as numbers—Betsy Jane, Pansy, The Coffee Pot, and the Rattler. Some locomotives had saddle tanks which carried water over the boiler. This added weight over the driving wheels and dispensed with the water tender. Then Ephraim E. Shay revolutionized the logging industry by construction of his lopsided geared locomotive. This machine had the boiler on one side and the cylinders on the other side where power was transmitted to the wheels by a system of gears and driveshafts. Because of the universal joints in its center drive shaft the Shay could negotiate tight curves; it also eliminated the track damaging effect of rod drivers connected directly to the wheels on conventional steam locomotives. A total of 2,761 Shays were built between 1880 and 1945.

The last one I ever saw in operation was in the Starkey area in the Blue Mountains in eastern Oregon in 1954. Following the introduction of the Shay and other locomotives of similar design, manufacturers improved on the design and efficiency of the gears. Electrified lines in the flat lands in Oregon came next, and finally the diesel locomotive. Then the romance of the "Coffee Grinder," "The Ant," "The Grasshopper," and the "Cyclone" died. So did the donkey engine. Now logs are brought out of the woods with balloons attached to giant trucks.

The modern conquest of the mountains has centered around many resources including summer feed for livestock, minerals, timber, water, and recreation. Men have sought the physical and biological resources mostly for immediate benefit and with little regard for the quality of life for future generations or for the effect on the mountains themselves. The devastation of forests, soil, grasses, and natural beauty is the legacy of man's prodigal exploitation of these natural resources.

Livestock came to the western mountains from two directions. Cattle, horses, pigs, and goats were common along the Atlantic Coast as early as 1775. Gregorio Villalobos landed calves near Vera Cruz on the east coast of Mexico in 1521. Coronado brought cows and sheep into the Southwest in 1540. From the Pacific Coast, livestock from the Franciscan missions spread eastward and northward. When the tides of grazing animals met from east and west the depletion of the nation's grasslands resulted from too many animals, lack of management, and pure greed.

When grazing capacity of the grasslands and deserts was diminished by drought and misuse of the forage, the herds became nomadic and were driven into the mountains for summer forage. Soon they devastated the meadows and high elevation grasslands. An estimated five million head of sheep were driven by shepherds or sheep herders from the Great Valley in California over the Sierra Nevada passes in spring and back to the Valley in autumn. Later, on the Wasatch Plateau in Utah sheep denuded the vegetation so that the number of bands could be counted from the desert at Ephraim by the dust clouds that rose from the distant mountain tops. From this barren ground floods roared down the mountain bringing mud and great boulders into the valleys with disaster to towns and farm lands.

Now, after years of research by federal agencies and Land

Grant colleges and universities better management practices have been developed. Grazing rates, seasons of use, and herding practices are now regulated by federal agencies on national forests and the deserts in the public domain. In some localities, however, it will be years or even generations, before the grazing lands approach a semblance of their former condition and productivity.

Concurrent with the rise of livestock use of mountain forage was the great exploitation of mineral resources. Extraction of oil in the Appalachians began in Pennsylvania as the Civil War supplies of whale oil declined. Coal, another mineral resource, became important as a source of coke for steel mills, for heating homes, running locomotives, generating electricity, and the production of gas for street lighting by distillation of bituminous coal. Now strip mining for coal has left enormous areas in dev-

Forage for livestock is one of the valuable renewable resources of the mountains. When sheep are properly distributed and not allowed to graze too long in one place, they are not destructive to the vegetation. They do prefer forbs which tend to disappear under heavy grazing. The result is a change to almost pure grassland.

astated condition in Appalachia along with economic depression, deforested hillsides, and contaminated streams and lakes. Programs of rural renewal have not solved the problem.

Less destructive, but of considerable impact on the mountain environment in Appalachia, has been the extraction of mineral products other than coal. These include cobalt, sand, iron, manganese, molybdenum, salt, sulfur, tungsten, vanadium, and zinc. In the limestone belt of the Valley and Ridge Province in the Appalachian Highlands, cement has been a major product. And from New England to the Piedmont Province, slate, granite, marble, and other building stones have been important local industries for many years.

In the Rockies and Sierras mining laid the foundation for development of towns, cities, railroads, schools, and industries. But as happens with nonrenewable resources the boom busted. Money and control passed to Eastern ownership, the forests were cut, and the industry destroyed itself. Now there is a great new boom on the horizon of the Colorado Plateau—the plan to extract billions of gallons of petroleum and trillions of cubic feet of natural gas from oil shales along the Colorado River. If the plan materializes, new towns, roads, schools, and extraction plants will be built. Taxes will rise, more police protection will be needed, water to operate the extraction process will be required, the greatest migratory deer herd in America will be displaced or eliminated, air pollution will increase, and the peaceful community life of the present residents will vanish. Where will it end? America needs oil!

America also needs water. The Northwest needs it for hydroelectric power. So does New York on the Hudson. So does Los Angeles, and Denver, and the Salt River Valley in Arizona. Soon the shale oil industry will want it in western Colorado. This water comes from the mountains. Will there be enough? Already the cloud seeders are experimenting with methods for increasing rainfall and the snow pack over the mountains. In the meantime we must save the water we already have.

There used to be a sign in a men's room in Canyon City, Colorado, on the Arkansas River that read, "Please flush the toilet. Pueblo, downstream, needs the water." This sort of thing can be remedied by adequate sewage treatment. But can hot water if it becomes radioactive from an atomic power plant be made clean

Mountain reservoirs store water for cities and for farm irrigation. As the water is drawn down in summer wave action makes terraces that tell a story of the rate of use.

so it does not poison people as well as boil fish? The new power plant on the St. Vrain River north of Denver is designed to be cooled by helium. Maybe we can see other ways to help save our rivers. It is up to each of us to get acquainted with our rivers and then see that our neighbors, our state and federal officials, and our congressmen also understand them. If we can prevent the desecration of our rivers the mountains will supply them with clean water.

The mountains now are becoming the playground for Americans. Too many of us in the 20th century have lived in the miasmatic world of the cities with their jammed freeways, air pollution, noise, high speed, high-tension jobs, and vicarious experiences by way of the TV tube. Not all of us have chosen this way of life; it has grown up around us, or we have been forced into it. Now we are finding that it is not enough.

Our interest is growing in another world, the world of the outdoors, the world of nature, where momentarily at least we can shed the burden of chaotic existence and visit primitive and restful surroundings. Call it outdoor recreation if you like, but by whatever name it includes many things. Its pursuits include fishing, hunting, hiking, skiing, camping, bird watching, boating, nature study, swimming, picnicking, horseback riding, wilderness touring, mountaineering, and photographing. For the exercise of many of these pursuits the appeal of the mountains is great.

Part of the appeal derives from sheer love of the outdoors, a change of scenery, and the opportunity to acquire an intimate knowledge of some facet of nature whether it be the structure of forests, streams, or mountains. Some people enjoy being authorities on small things even if they are nothing more than ferns, orchids, warblers, or minerals. Frequently these small but furious pursuits lead one into a second career. I know a man who started collecting rocks and now in retirement he cannot supply the demand for his fine jewelry made with gemstones. Other people like to establish records or belong to select groups. The Adirondack 46ers are a mountain climbing group with the membership requirement that one must climb all forty-six Adirondack peaks judged by an 1897 survey to be 4,000 feet or higher. Now a few are suggesting that membership be limited to 1,000 since some people are becoming "peak grabbers" who climb the forty-six peaks only as a stunt and without regard to littering, erosion from foot paths, destruction of wildflowers, trail maintenance, or the occasional need for rescue teams. Another type is the jet boater who likes the power, the thrust, and the challenge of the rapids on wild rivers such as the Snake River in Hells Canyon. Jet boats are fun for the riders but nothing intrudes on the wilderness like the sound of an unmuffled motor.

Competition of various kinds provides vicarious enjoyment for people who visit the mountains for only a day. In many places in the West there are rock drilling, log sawing, and tobacco spitting contests. These may be followed by four-wheel drive vehicle rallies and greased pig races. In the annual festivities at Leadville, Colorado, the annual Boom Days Burro Race attracts both men and women. The contest is a twenty-one mile race with a burro to the top of 13,000 foot Mosquito Pass.

In contrast with these short term trips to the mountains is the unforgettable experience of a visit to a great wilderness area. One of these is the Idaho and Salmon River Breaks Primitive Area and adjacent wild lands which encompass nearly two and a half million acres of some of the most remote, pristine mountain country on the continent. In this yet unspoiled primeval

Author with salmon caught on the Hoh River in the Olympic
Mountains. Anadromous fish come up from the Pacific Ocean to spawn
in the cold glacial water that flows down from Mount Olympus.

wilderness are big horn sheep, mountain goat, elk, deer, mountain lion, black bear, fisher, otter, coyote, lynx, bald and golden
eagles, osprey, anadromous fish, and native trout. The rivers are
wild and scenic and the forests and wildflowers provide a place
of bliss for botanists. Ultimately this wild region will need protection too. In 1972 the Forest Service recorded 257,000 visitor-
days of use which included boat trips, hunting and fishing activities, and sightseeing trips with pack horses. It is estimated that
the area will receive one and a half million days of visitor use by
the year 2000.

In the mountains are many values all of which in proper per-
spective are necessary to the existence of the nation. The multiple use concept, which proposes that management integrate all
resources into utility for society, is fine in theory but self-
contradictory in practice. A campground built in a stump patch
after the trees have been harvested may be an example of multiple used in name but it conflicts with the receptionists' interest.
Several uses on a single area cannot always be complementary.
On the other hand, true single use is impossible since a forest,
harvested or not, or a wilderness area, used or not, is still a home

for wildlife and a producer of water which redounds to the benefit of the whole public.

A great need in the management of mountain resources is to view their use, not as development but as exploitation. Human "development" seldom improves on nature; instead, it induces irreversible changes, and seldom benefits the whole public interest. Good management also requires reliable information so the public can participate in the formation of goals and priorities to assure that their interests are being served. Pressure groups, self-centered government agencies, and opportunistic industries should not have the prerogative of telling the public what its interests are.

Preservation of the environment requires the skill of many. Proper planning for a timber harvest, for example, involves the skills of forestry technicians, civil engineers, soil scientists, wildlife specialists, and hydrogologists. Coordination of timber harvest with other uses of the land also should involve the skill and knowledge of landscape architects, environmentalists, and representatives of the general public. This is being done to some extent by some of the public land managers. Environmental education centers with programs concentrated on teachers and students are providing some of the public with a new awareness of where man stands in relation to his environment. Furthermore, we must be eternally vigilant to detect fresh threats to the resources. Some of these threats seem to appear spontaneously.

One of these, a newcomer in the mountain environment, is the snowmobile. The booming popularity of this recreation vehicle, like the jeeps, pickups, dune buggies, and trail bikes, is creating an ever increasing amount of vandalism and environmental damage across the country. Four million or more of these destructive noisemakers are roaring about the country.

These machines with engines up to ninety horsepower and speeds of sixty to eighty miles per hour are capable of penetrating remote mountain areas with ease. They permit quick escape by vandals of unoccupied mountain cabins in winter. They contribute to littering of trails, erosion, destruction of young trees, and are a menace to wildlife. Even if deer and other animals are not pursued to exhaustion in deep snow the noise drives them away from their food supply and may lead to starvation.

The noise also destroys the serenity of the woods for hikers, horseback riders, picnickers, and bird watchers. The ruts left in snow also are hazardous to skiers.

Various solutions have been recommended by The Conservation Foundation: zoning of public lands to prevent entry of off-road vehicles to ecologically delicate areas; registration and taxation of these vehicles to provide funds for enforcement of regulations concerning their use; and research on reduction of noise, on air pollution from oil-gasoline mixtures, and on the danger to fish and other wildlife. The most difficult problem is to educate the users to an appreciation of the need to minimize detrimental effects to other people and to the environment. In the final analysis the users of these vehicles are the real problems.

In a broader sense management of the recreation resource requires more reliable information about the user public as well as the resource itself. People's quality perception of the outdoor environment is largely directed by the urban syndrome. People who leave the city for the outdoors are conditioned psychologically to take their city environment along with them. Their motivation is to live for a time in new surroundings but with all the conveniences and accouterments of home, including TV, flush toilets, electric lights, and propane heated trailers.

This camper profile has its redemptive features, especially where private and public agencies supply home-type facilities. People are inclined to stick close to services and thus produce smaller impact on the natural environment. When the environment is close by, as in National Parks, the impact is reduced by requiring people to stay on designated trails or by reducing the number of camping units per area and requiring advance registration for occupation of camping sites.

A different breed of outdoorsman are the hikers and trail users. Many of these are confirmed conservationists who love exercise, seek esthetic values and simple contact with nature. Their presence in the wilderness of the mountains is inconspicuous and consequently some managerial personnel consider them to be a nonvocal minority and would like to relegate them to semiwild areas, settled regions near urban centers, and even city parks. Thus the expense of maintaining trails in the mountains and on public lands would be lessened.

The trail system, however, is well established in the East and is receiving more attention in the West. The Appalachian Trail, from Mount Katahdin in Maine to Springer Mountain in Georgia has set the precedent for all other trails. Its hikers are dedicated people, and true conservationists in their efforts to maintain the trail and its environment. Excitement, adventure, discovery, and refreshment of the spirit greet the hiker on this simple footpath through the wilderness. Portions of the trail are near enough to civilization that anyone can walk through a part of its natural beauty.

The Continental Divide Trail from the Canadian border in Montana to the desert mountains in New Mexico is largely undeveloped but extensions, such as the Escalante Trail in Colorado and Utah, are being planned. The Pacific Crest Trail from northern Washington to Baja California already receives much use, especially in the Sierra Nevada. The trail concept now is promoting study of bicycle and horseback trails, metropolitan trails, and historic trails such as the Oregon Trail, the Lewis and Clark Trail, and Potomac Heritage Trail. Parts of these will traverse majestic mountain scenery. Their greatest benefit will be to get people to see close at hand the heritage our mountains have given us and to understand that the mountains have their part in directing the destiny of us all.

You can enjoy the outdoors with a minimum of danger and a great deal of pleasure by living in a mountain cabin for a few days or even weeks. From this sanctuary you can make short trips to rivers, forests, and geological formations, and can study plant and animal life at leisure. Your visit will give you an intimate knowledge of the local mountains and their ever changing moods. The experience also will give you confidence in your ability to cope with nature. And your growing familiarity with principles, processes and things in the wild will make you environmentally oriented.

My wife and I used to spend a few days at different seasons in a log cabin in the Oregon Cascades. The place was sparsely furnished with rustic furniture which matched the mood of the surrounding country. A monstrous thing, known as the "Honeymooner's Bed" was made of pine posts and crosspieces slotted into the uprights—it must have weighed half a ton. The high spring and still higher mattress, with a sag in the middle, could kill your sacroiliac. But on çold autumn nights, with the tea-kettle still singing on the kitchen stove you could sleep like Somnus.

The cabin was interesting in other ways. On the heavy pine table the impedimenta of previous visitors, who planned to return later, were always there—fishing nets, fly hooks, an alarm clock, a roll of socks, a book on music, and maybe a flashlight. Usually some thoughtful person had laid the fire in the fireplace and brought kindling for the kitchen stove. Everything else was neat and orderly. But the lock on the front door did not work. You had to kick hard or bang the thing with your fist before the lock would function. The noise would scare the skinny chipmunk on the stoop and he would dash through the frosty forest litter for his hole beneath a ponderosa pine log.

After a moonlit night the day would start crisp, maybe at 15° F., and warm rapidly as the sun dispelled the fog that shrouded the river. The jays would come alive and the trout would start jumping as insects became active along the river banks. Then we would take long walks in the pine woods or drive to nearby lakes, or to the shining volcanic mountains, or explore lava caves. And in the evening we would go down the river and fish near a waterfall.

Sometimes a gray shadow would float up from the deep and suck in my renegade fly and then submerge as I set the hook. Once in the growing darkness I thought I had connected with the river bottom, but it moved. It never did come up until my net lifted a seventeen-inch brown trout. We ate two of them, topped with white wine and a dash of lemon. And then the fire in the fireplace burned brightly.

Modern people are largely unaware of the pleasure a radiant fire brings in a mountain cabin on a cool October night. We are so accustomed to the adulterated secondhand warmth we get from our intermittent blowers, gas furnaces, and low-humidity houses we have forgotten or never experienced the reality of fire itself. Fire on an open hearth warms you to the bones. When you stoop to straighten the log the shower of sparks is a pyrotechnical reminder that you are in the primitive again and the hot burn of your scorching pants when you stand with your back to the glowing embers sends shivers of pure pleasure up your spine.

Once the walls and floors of your cabin are warm you radiate less of your own heat to the surroundings. Then you need less fire and can even let the flames subside and the glowing embers bring nostalgic memories of other carefree days. This pleasant surcease from the demands of everyday living is one of the gifts of mountains.

Backpacking in the wilderness attracts many people. Here there are no automobiles, four-wheel drive vehicles, air pollution or TV programs. Only the woods, the mammals, the birds, the wildflowers, and nature relatively undisturbed by man are there to enjoy. This is the Three Sisters Wilderness in central Oregon.

Appendix

How to See the Mountains

WITH REASONABLE TIME AND EFFORT it is possible to see and experience the awesomeness and grandeur of the mountains from almost any point in the United States. From New England to Georgia and westward to Arkansas, Oklahoma, and southern Missouri one can reach magnificent mountains and uplands by car or auto camper. Even for people who live in the midlands of America, the Wichita Mountains in western Oklahoma, the desert mountains in western Texas, and the Black Hills in South Dakota are easily accessible. From the Rocky Mountains to the Pacific Coast, mountains are hardly ever out of sight and the outdoor enthusiast can reach them with only a few hours of travel.

Most of the mountains are in public ownership of one form or another. Much of the National Forest ownership is characterized by mountain topography. Millions of acres in the West, including desert mountains and forested mountains along the Pacific Coast are supervised by the Bureau of Land Management. Much of the spectacular scenery of the nation now is protected by the Park Service of the Department of Interior. Many other outstanding areas where scenic geology, wildlife, and recreation are available are maintained as state parks, wildlife refuges, monuments, and historical sites. In all of these areas, subject to regulations of the managing agencies, the interested visitor may go to see nature in action and to enjoy mountain adventure.

The way to see mountains is not to drive over paved highways with the landscape flitting by and the radio playing rock music. To really see the mountains you have to experience them physically, mentally, and aesthetically. You have to walk among their forests, clamber over boulders, and claw your way up to the summits. You also have to pause and leisurely become acquainted with rock structures, plant life, and animal inhabitants. Only by becoming intimate with your surroundings and experiencing their daily, monthly, and yearly changes will you truly appreciate the complexity of the mountain environment and the reality of the earth beneath your feet.

Mountain adventure takes many forms and provides variety for any taste. It offers physical challenge to hikers with all degrees of competence, from those who want to sample the fragrance of wildflowers and feel the soft earth of meadows beneath their feet to those who pit themselves against massive cliffs and the elements and view the conquered mountain as their reward. It offers compensation to the artist and photographer in the form of compelling subjects which capture the spirit and appearance of unique land forms and natural objects. It offers natural beauty, the song of birds, and solitude for meditation.

To appreciate mountains it is best to travel leisurely. If you have only a day, decide on a few points of interest and sample them thoroughly. Relax beneath a tree on the banks of a dashing stream and enjoy the antics of a dipper or water ousel as it bobs on a stone and then walks beneath the water in search of insects and then appears clean and dry to bob again on its favorite stone. Spend an hour on a hand and knee exploration of the diminutive wildflowers on a room-sized area of alpine meadow. Or explore a canyon with its brook, its streamside flowers, its swallow nests on rock walls, and the contrasting trees and shrubs on its opposing shadowed and sunlit sides.

A day in the mountain world is not enough. A week of intimate exploration around a favorite campground will acquaint you with some of the common plants, birds, mammals, and local vistas. But this is only a beginning. In a month all will be changed. New plants, new birds, new animals, and different weather conditions will give the surroundings and the whole area a different aspect with new colors and different vistas.

There is no best season to see the mountains either. Some prefer the freshness of spring when green things spring from the earth, when birds return, and when animals bring forth their young. Others enjoy the fullness of summer when verdure covers the land, the sunshine is warm in the daytime, and transient storms bring refreshing rain. The fall mood is one of riotous color, falling leaves, and invigorating winds that bring the threat of snow and nippy nights. Then the wildflowers fade and disseminate their seeds, the small mammals hibernate while the larger ones migrate to lower altitudes, and the migratory birds vacate their homes for southern climes. Then winter comes with exhilarating days for skiers and experienced adventurers who are willing to risk the blizzards that turn the mountains into a snowy fastness that has its own frigid beauty.

To find the tranquil places, you must get away from surfaced roads and commercialized tourist resorts. Follow the back roads and then walk from there. You will have to go only a short way into the hills to be free of the crowds since most people do not travel far from their means of transportation. And you may find that walking is an invigorating experience.

TRAIL TRAVEL

There are thousands of miles of foot trails in the mountains of the United States. These are the footpaths to wilderness and the true flavor of the outdoors. A visit by automobile can give you views from the lowlands or from the heights. The ride requires little physical exertion and the swiftness of the journey provides fleeting vistas. But the true meaning of topography, the real essence of the environment, and the enjoyment of forests, birds, mammals, wildflowers, weather, and the soil beneath your feet can best be achieved only by walking and savoring the mountain world with all your human senses.

You can set out on short walks in any of the National Forests, National Parks, state forests and parks, and on a multitude of lesser recreation trails. Or you can spend weeks or months on the mighty trail systems that traverse mountains for distances of 2,000 miles or more. The long distance ones are the Appalachian and the Pacific Crest National Scenic Trails. The Continental Divide Scenic Trail, still existing primarily in name only, will follow the backbone of the continent through the Rocky Mountains. At least one man, Eric Ryback, has already walked from Canada to Mexico near or along this divide from which the waters of the nation flow west and east.

The Appalachian Trail is the preeminent one, 2,050 miles of continuous footpath, extending from Katahdin's granite heights in Maine to Springer Mountain in Georgia. It is the pioneer of all foot trails, the best developed, and the best marked. Guide books are available for planning trips and following the trail. Huts and open shelters, more than 230 of them, now are available for accommodation at night and in stormy weather.

The Appalachian Trail passes through fourteen states and traverses wilderness, high mountain ridges, primeval forests, mighty rivers, including the Hudson, and scenic valleys. From the Katahdin wilderness in Maine the trail crosses the White Mountains in New Hampshire and runs southward along the Green Mountains in Vermont and the Berkshire and Taconic Mountains in Massachusetts and Connecticut. Then it crosses a series of ridges on the way to the Kittatinny Mountains, the Alleghenies and then the Blue Ridge. Passing through the Great Smokies the trail leads through the Nantahala Mountains and terminates at Springer Mountain in Georgia. The Appalachian Trail is a foot trail and is not adapted for horse or bicycle travel. Motor vehicles are generally prohibited by Federal law.

The proposed Continental Divide National Scenic Trail will take the hiker down the backbone of the North American continent through some of the nation's most prominent topographic features. Since the Rocky Mountains do not form a continuous chain of ridges and peaks the trail will pass precipitous peaks, sparkling lakes, alpine meadows, verdant forests, and semidesert basins between mountain ranges. The 3,100 mile route will afford views of some fifty peaks over 14,000 feet. In passing through five western states the trail enthusiast will encounter five ecological life zones with their enormous variety of plants, animals, soils, slopes, exposures, and climates.

The Continental Divide Trail will begin its first segment of 1,235 miles at the Canadian border in Glacier National Park amid glacial lakes, rocky cirques, alpine vegetation, and coniferous forests. Much of this high mountain land will be accessible on foot or horseback. Leaving Glacier National Park the trail will continue southward through national forests to the Bob Marshall Wilderness where one hundred miles of trail already exist. Continuing southward, the trail will wander back and forth across the Divide in Montana and Idaho to Yellowstone National Park, on to the Wind River Mountains where 13,804-foot Gannett Peak is the highest point in Wyoming. The segment will end at South Pass City where the Oregon Trail once crossed the Continental Divide.

The South Pass City to Medicine Bow, Wyoming, segment of the trail will traverse desertlike country. Here the Divide splits to form the east and west rims of the Great Divide Basin with internal drainage where water flows to neither the Atlantic nor the Pacific Oceans. This is jade and petrified wood country and the ancestral home of the pronghorn antelope and the American bison. Wild horses still run in the basin. Pinnacles and buttes rise above the sagebrush-grassland in this almost uninhabited land.

From the Medicine Bow National Forest in Wyoming to the Jicarilla Apache Indian Reservation in New Mexico the proposed trail will range in elevation from 7,500 to 13,000 feet through forested mountains and the nation's most extensive arctic alpine life zone. Recreationists on this trail will be able to indulge in fishing, hunting, climbing, photographing, snowshoeing, artistry, and study of geology, ghost towns, mines, abandoned railroads, and livestock grazing. The trail will pass through many of the mighty mountain ranges in Colorado, including the Front Range, the Sawatch Range, and the incomparable San Juan Range.

In Colorado there are plans for a trail system to connect with the Continental Divide Scenic Trail. This fledgling system will be for everybody—hikers, bicyclists, horseback riders, and even blind and handicapped persons. The basic trail network will begin close to the major cities from which people can go to the mountains for temporary escape from the humdrum of daily living.

From the Jicarilla Apache Indian Reservation to Silver City, New Mexico, the Continental Divide Scenic Trail will pass through a variety of topography ranging from mountainous and canyon country to plateaus and mesas in semidesert lowlands. The mountainous terrain is characterized by ponderosa pine forests. Geological forms include ancient volcanoes and

Overnight camping trips in the mountains provide an ideal way for families to enjoy solitude and become acquainted with nature.

lava flows. From Silver City, New Mexico, the trail is projected to continue southward through semidesert grasslands and various mountain ranges, including the Pyramid Mountains and the Animas Mountains, to end at the United States-Mexico border. The entire trail, from Canada to Mexico, when completed, will provide a magnificent travel experience in the high mountain world of America.

The Pacific Crest National Scenic Trail was officially recognized by the National Trails System Act of October 2, 1968. When completed the trail will be some 2,350 miles long and will be located mainly along the mountain ranges of Washington, Oregon, and California. Previously developed segments of this trail include the Cascade Crest Trail in Washington, the Oregon Skyline Trail along the Cascades, and the John Muir Trail in the Sierrà Nevada in California.

The Washington portion of the Trail follows the backbone of the Cascades from the Canadian border to the Columbia River, a distance of more than 450 miles. Walking with a backpack requires about one month. Beginning at Monument #78,

United States-Canada border, the trail passes through Glacier Peak Wilderness, North Cascades National Park, Lake Chelan National Recreation Area, and many high mountain passes. The trail follows the eastern boundary of Mount Rainier National Park for eleven miles, passes through the Mount Adams Wilderness and ends at the Columbia River at the Crest Trail Inn east of Vancouver, Washington. The crossing into Oregon can be made over the "Bridge of the Gods" or at the Hood River Bridge over the Columbia.

In Oregon the Trail follows the skyline of the Cascades past Mount Hood, Mount Jefferson, Three-Fingered Jack, Mount Washington, the Three Sisters, and the lake region, including Waldo, Crescent, and Diamond Lakes. The Trail passes through Crater Lake National Park where the 2,000 foot deep lake lies in the bowl of the extinct volcano, Mount Mazama. South of beautiful Lake of the Woods the Trail passes into California.

The California section passes through fourteen national forests, four national parks, five state parks and various private lands. Elevations vary from 500 feet to 13,200 feet and 94 peaks of over 13,000 feet may be seen, including 14,496-foot Mount Whitney, the highest peak in the conterminous United States. The area traversed includes all vegetation zones from semidesert to arctic alpine. Magnificent spruce, pine, fir and sequoias may be seen in the high Sierras. Fauna in the various zones ranges from salamanders, chipmunks, and marmots to deer and bears. Birdlife is abundant. Topographic, geologic and cultural features range from high peaks, fault-block mountains, lakes, glaciers, ice caves, and ghost towns to gold mines. The Trail in California is not complete but temporary travel routes are being identified. Many side trails now connect with the established sections of the main travel route.

Various regulations for use of trails have been established to preserve natural conditions in wilderness areas, national forests, national parks, and other ownerships. Permits are required for entry into many of these areas and special regulations are in effect for certain kinds of use and also during fire seasons.

In wilderness and national park areas where pack animals are permitted there are rules for handling, hobbling, picketing, feeding of horses and their management on trails. Horses can be destructive to fragile environments, especially if they are allowed to trample vegetation in campgrounds, pollute water supplies, graze in one place when tethered, and roam over wet spongy meadows.

National parks impose the requirement of "pack in–pack out" for paper, tin cans, and other litter. No guns, dogs, or cats—even on a leash—are allowed since these disturb wildlife and are a nuisance to visitors who wish to appreciate the undisturbed environment. Backpackers and campers now are asked to use portable gasoline or propane stoves since use of wood from standing dead trees destroys a form of beauty that is a part of the scenery and solitude of the mountains. Even where the use of dead wood which has fallen to the ground is permitted, it should be used sparingly since it is a part of natural recycling and its replacement is a slow environmental process.

On national forest lands special regulations apply to use of campgrounds, water supplies, and care of camp fires. During critically dry seasons camp fires may be prohibited altogether. It is recommended that truck-campers and automobiles be equipped at all times with an axe, shovel, and water bucket for use in extinguishing fires. Camp fires should never be left unattended. In wilderness areas, on scenic trails, and on many other trails and fragile environments use of motor equipment is forbidden. These include four-wheel drive vehicles, scooters, motorcycles, and snowmobiles. These are destructive to trails, vegetation, and wildlife. Also, they are sound polluters in an environment where the true outdoors person seeks peaceful solitude and unspoiled beauty in a world that is constantly growing smaller.

Mountain Survival

There are hazards in the mountains. Almost daily, reports in the media tell of someone, somewhere in the mountains of the United States being lost, injured, drowned, fallen from rocks, suffering from exposure, or being involved in automobile accidents. There are other natural dangers in the mountains. Heart attacks occur when people not in good physical condition attempt to climb mountains too rapidly. There are occasional snake bites and in the western mountains ticks can inoculate one with spotted fever. And in desert mountains heat exhaustion and thirst are real dangers for the person who goes unprepared.

These hazards do not mean that mountain travel is more dangerous than travel on the paved highways; or existence in the big cities where theft, mugging, murder, and confusion are always present; or even in the home where accidents, fire, poison, and other dangers take their toll. Considering the millions of people who enjoy the outdoors there are far fewer accidents and fatalities in the mountains than in the world of our daily lives. With reasonable knowledge of what may be mountain hazards, with precautions to avoid danger, and with adequate clothing and equipment the mountain journey can be a safe and enjoyable experience.

If you are going into the mountains, at least a part of your trip will be by automobile, pick-up camper, or other motorized vehicle. Your vehicle should be in good condition, including the engine, tires, and brakes. On paved mountain roads observe the posted speed limits but do not crawl at extremely low speeds. This causes engine overheating, and is hazardous and exasperating for other motorists, especially those climbing the grade behind you.

Honking the horn at every curve, as some inexperienced or overly cautious drivers do, is useless and is a distraction for the driver and other occupants of the car. Stay on your side of the road, even if the precipice at the side of the pavement seems to drop away to eternity, and trust that the driver coming from the other direction will do the same. If your engine overheats, stop at the first safe turnout and let the motor idle at one-fourth or one-third throttle until it cools. If you turn off the ignition too soon the radiator is sure to boil over. If it boils do not check the water level until it cools; otherwise you may be burned by steam.

In descending steep mountain grades use the lowest gear position that will allow the engine to act as a brake. Shift into the proper gear position, especially if the "low-low" gear is needed, before the car attains a high downhill speed. If you maintain the proper gear position only a minimum use of brakes may be necessary. Long steady pressure on the brakes may cause them to "fade." If the brakes do begin to fade, stop at the nearest safe turnout and let them cool for at least twenty minutes. Remember that cars going up the mountain should have the right of way over cars going down.

You should never drive on secondary roads in the mountains without tire chains, an axe, shovel and tools for minor repairs. On poorly maintained dirt roads mud holes are common and chains may be necessary to negotiate these hazards. Snow tires on passenger cars are not as reliable as chains when deep mud is encountered. If ruts are deep or rocks are present your vehicle may become high-centered. The axe can then be used to cut dead wood to build a platform for the bumper jack to extricate yourself from a difficult situation. Extra oil, water, and a fan belt are desirable accessories when traveling or camping several miles away from main highways. Check your motor before starting in the morning. A pack rat may have built a stick nest around the fan belt during the night. Or a mouse may have constructed a home of flammable material on top of the carburetor.

A map and compass are important accessories if you plan to walk from car or camp in unfamiliar country. Learn to use them before you start. The topographic maps of the U. S. Geological Survey are useful if you travel in rugged terrain since they show physical features, roads, streams, and places of human habitation. These maps may be obtained at some outdoor stores or from the Map Distribution Office, U. S. Geological Survey, 1200 South Eads Street, Arlington, Virginia 22202. Use your map and compass to orient yourself with visible features—hills, peaks, rivers, and distances. As you travel look back to familiarize yourself with the appearance of the country you have already traversed so it will not seem strange to you if you have to retrace your steps.

For travel on the Appalachian Trail and connecting trails, information about guide books and maps can be obtained from The Appalachian Trail Conference, P.O. Box 236, Harpers Ferry, West Virginia 25425. For travel in the western mountains recreation maps may be obtained from local offices of the U.S. Forest Service and the National Park Service. These publications show trails and roads and give information on regulations concerning travel, camping, and safety precautions.

If you plan to walk far into the wilderness remember that communication with the outside world is limited. Plan to be self-sufficient for a longer time than your journey would ordinarily require. The veteran hiker, when going into the back country, carries a map and compass, matches in a waterproof container, a sturdy knife, insect repellent, sleeping bag, clothes adaptable to weather changes, and emergency food. He or she keeps the weight down by carrying only essential aluminum cooking utensils and powdered, dehydrated, or freeze-dried foods. And the hiker takes a pencil and paper for writing notes or for leaving messages if he becomes lost.

The clothing you carry is important even if you travel in summer. Mountain weather is extremely changeable. The temperature can drop one degree per minute and the wind can rise abruptly from calm to a gale. This introduces the chill factor, the rate of heat loss from exposed skin. A wind speed of thirty miles per hour with a thermometer reading of 0° F. can be equivalent to 50° F. below zero. Add clothes *before* you become chilled.

Long pants are advisable for extended walking trips in the mountains. They protect against cold, sunburn, prickly bushes, rock scrapes, and annoying insects. On high altitude summits and in rain country wool pants are best. Light weight down-filled jackets with water repellent material are easy to carry. A poncho or small polyethylene tarp which can be slipped over the head is better than a rain suit which collects body moisture and makes you stew in your own sweat.

If the lure of the trail takes you into the high country of the West or into the desert mountains you should give some thought to the effects of altitude and dehydration. Don't ruin your hike by walking too fast or by strenuous rock climbing when you are not acclimated to the lack of oxygen at high altitudes. Use sunglasses, especially if you are on snow, and carry suntan lotion to avoid sunburn even on cloudy days. Adhesive bandages should be applied to rubbed spots on your feet *before* blisters appear.

In dry desert mountains thirst is the first symptom of dehydration. Other symptoms that follow are nausea, dizziness, difficult breathing, and heart pounding. Dehydration of fifteen to twenty percent can be fatal. In these mountains where high temperatures, low humidity, intense sunlight, and brisk winds are common, carry water. Use salt, but use it only with water, since salt alone increases the rate of dehydration. Keep clothes on to reduce sweat evaporation. And remember that even desert mountains are subject to enormous temperature changes from day to night.

The terrain in desert mountains also is different from that in forested mountains. Rains are infrequent and rocks are loose and may crumble and roll. Gravel slides are unstable. And the foliage of cactus or yuccas with their needle-pointed leaves are unfriendly. Canyons, hills, and mountain ranges tend to look alike to strangers—be aware of your surroundings, know where you have been, and know where you are going. With these precautions in mind, enjoy the desert mountains. They are different. Their life zones are conspicuous and their plant and animal life is fascinating. Most can be climbed by easy trails, by scrambling up ridges, usually with a minimum of rock climbing skill. Peaks in true desert mountains are barren of trees but the summits of other mountains that rise out of the desert are forested oases with sparkling lakes and flower strewn meadows.

There are snakes in the mountains. But then there was a snake in the Garden of Eden. We have to accept the presence of these and other pests just as we accept the greater dangers in our daily lives. Relatively few people ever see snakes in the mountains but a bite by a poisonous species is serious and medical attention should be sought immediately.

Among the less dangerous snakes in the mountains are the pigmy rattlesnakes and the copperheads. The pigmy rattlesnake or eastern massasauga (*Sistrurus catenatus catenatus*), a medium sized serpent with rattles of medium size, occurs from central New York and Pennsylvania to Missouri. The northern copperhead (*Akistrodon contortrix mokeson*) is a moderately heavy bodied snake with chestnut color on top of its large triangular head and an upper body with reddish brown cross bands on a ground color of light reddish tan. Its bite is serious and extremely painful but seldom fatal.

Copperheads range from the Ouachita and Ozark Plateaus in Arkansas, along the Appalachian Highlands, and into Pennsylvania and southeastern New York. The massasauga and the copperheads prefer rocky wooded hillsides and use limestone ledges for dens. They also live around sawdust piles because of the numerous mice that breed there. Watch where you step and where you place your hands. Stir leaves with a stick before you pick up stones, flowers or mushrooms. Because of their protective coloration copperheads are hard to see among dead leaves. The western cottonmouth or cottonmouth water moccasin (*Akistrodon picivorus leucostoma*) is a heavy bodied snake with a large triangular head and eyes covered by supraocular scales so that the eyes are not visible when viewed from above. By contrast, the eyes of common watersnakes are located high on the head and are clearly visible as one looks down at the snake. The western cottonmouth is not a typical mountain snake but it is common in ponds, drainage ditches and in clear rock-bed streams in the Ozark region. Its bite is more serious than that of the copperhead.

The true rattlesnakes are common throughout much of the United States. The timber rattlesnake (*Crotalus horridus horridus*), a moderately large snake, is found with copperheads from the Ozark Highlands, through Kentucky and Tennessee, in the Appalachian Highlands, and northward to New York and Maine. In the West, the large western diamondback rattlesnake (*C. atrox*) and the Mohave rattlesnake (*C. scutulatus*) are found in mountain foothill country from west Texas to the Mohave Desert in California.

I have encountered various varieties of the western rattlesnake (*Crotalus viridis*) in the canyons in the Rocky Mountains, in the Great Basin, and in the northern Cascades in Washington. In all, about twenty-five species of rattlesnakes are recognized. In general they are not common in mountainous country. All have fascinating life histories, are the subjects of vivid folklore, and are important in nature's food chain. Some are bad tempered and all should be avoided by keeping a watchful eye when traveling on foot. High top boots instead of ankle-length shoes are good protection when you are in snake country.

In case of snake bite a snake-bite kit should be used and its directions followed carefully. The first step is to apply a constricting band above the fang marks but not so tightly that arterial circulation is restricted. Directions in the snake-bite kit will tell you how to make shallow incisions above the bite and how to use suction cups or your mouth to remove the venom. Treatment by a physician, of course, should be sought as soon as it is physically possible.

Although snakes and other creatures in the mountains occa-

Pancakes and trout fried on a sheepherder stove add to the pleasure of wilderness exploration. This remote camp is on the Ashnola River in the Northern Cascades near the Canadian border.

sionally present problems, a greater danger occurs when you become lost. Then you need self-control and self-reliance. People commonly get lost because they are attracted away from camp or trail by birds, animals, wildflowers, or just the enthusiasm of walking without observing their surroundings. Side roads, logged areas, thickets, and even unmarked trails can lead one astray. The seriousness of being lost comes when you are caught in a storm, fog, or snow, or have to stay overnight without adequate food, clothing, or shelter.

In high rugged western mountains or in regions subject to dramatic weather changes, such as occur on Mount Washington in New Hampshire, mountain travel can be extremely perilous.

When you fail to appear at a predetermined time a search by competent woodsmen is required. This may involve officers of the U. S. Forest Service, Park Service, State Police, Sheriff's Office, airplanes, army helicopters, and special rescue units. Their search and rescue will be made easier if you let friends and management agencies know where you plan to go before you attempt any extended journey into the back country.

If you are lost, trust your map and compass, if you have them, to indicate the direction and distance to the nearest trail, road or stream. To find your present position climb to high points from where you can see the surrounding country and identify topographic features on your map. Follow telephone lines if you find them. Usually they will lead you to houses, ranger stations or main roads. In the western mountains do not necessarily go downhill. Rivers in the valleys, of course, will eventually lead you to human habitations but many mountain trails follow the ridges instead of the valleys in remote wilderness areas. I once helped in the search for a hunter on the rim of Hells Canyon above the Snake River. If she had gone down hill over precipitous slopes she would have found no roads and would have been many miles from civilization.

Experienced hikers carry a small mirror which can be used to flash signals to searchers in aircraft. They also know how to build fires and make smoke with green boughs placed on the fire. If they believe search and rescue teams may be looking for them they remain in one place, preferably in open meadows or treeless areas, instead of wasting energy dashing aimlessly through the woods. They do not travel at night. Instead, they rig emergency shelter with evergreen boughs or with their poncho or plastic tarp. In country with deep snow they construct a snow cave— a narrow trench some three feet deep, covered with branches and then snow. The end of the trench should be kept open for adequate ventilation. A sleeping bag covered with snow can be quite comfortable. I know, since I have been covered with snow during the night in the Rockies and in the northern Cascades.

Glossary

ALPINE. The mountain region above upper timberline. Timberline is the upper edge of tree growth.

ANDESITE. A gray fine-grained volcanic rock consisting of feldspar and other minerals.

BALDS. Treeless areas on the mountain tops in the southern Appalachians. Heath balds are densely vegetated with shrubs belonging to the heath family. Grass balds are populated with grasses and sedges. Windfall of forests, landslides, and fire in former times are believed to be factors that explain the presence of balds in the upper tree zone.

BASALT. A dark fine-grained igneous rock containing feldspar and the black mineral pyroxene. Mostly of volcanic origin. Also found in dikes or intrusions between other rocks.

BATHOLITH. A mass of igneous rock, more than forty square miles in area, solidified from the molten state before reaching the earth's surface. The mass, consisting mostly of coarsely crystalline rock, extends to an unknown depth.

BOG. A fresh water aquatic habitat, high in concentration of humic acids with waterlogged peaty soils and unique vegetation consisting largely of sphagnum moss and plants of the heather family. Bogs commonly have a floating sedge mat which gradually encroaches on the lake and fills it with peat. Marshes have sedges, cattails, rushes, and reeds. Swamps have shrubs and trees along with plants characteristic of bogs and marshes.

BRACT. A modified leaf or scale, usually small, near the base of a flower cluster or inflorescence.

CALDERA. A large basin-shaped depression, wider than it is deep, caused by explosion or collapse around the center of a volcanic crater.

CALYX. The outer series of floral leaves or sepals, especially when they differ in color, size or shape from the inner series of petals.

CAMBIUM. The living layer of cells in stems and roots of plants that produces wood on the inside and food-conducting tissue and bark on the outside.

CARRYING CAPACITY. The productivity of a habitat or environment that will support an optimum number of consumer organisms without deterioration of soil or plant cover.

CATKIN. A spikelike inflorescence of unisexual flowers, usually pendulous. Willow flowers are examples.

CHLOROPHYLL. The green pigment in plants that manufactures food by photosynthesis.

CHRYSALID (OR CHRYSALIS). The firm case or pupa of a butterfly. The pupa finally develops into the mature insect by an internal transformation or metamorphosis.

CINDER CONE. A volcanic cone composed of cinder particles around a crater.

CIRQUE. A deep steep walled or amphitheaterlike depression, commonly semicircular, in a mountain caused by glacial erosion at the head of a valley.

CLIMAX COMMUNITY. The ultimate biotic community produced by succession and capable of replacing itself indefinitely. A beech-maple forest is an example.

COACTION. The interaction between organisms of different species which live in intimate association with one another. The relationship may result in benefit to only one organism (commensalism); mutual benefit to both organisms (mutualism); or, damage to one organism (parasitism).

COMPOSITE. A member of the thistle family. Flowers are borne in heads on a common receptacle surrounded by bracts. Examples are dandelions, sunflowers, asters, and goldenrods.

CONGLOMERATE. Sedimentary rock consisting of rounded rock and mineral particles cemented together, usually in a sandy matrix. The larger rocks and fragments may range in size from pebbles to boulders.

CONY. Another name for the pika or rock rabbit; it is related to the rabbits and hares.

CORM. A thickened vertical solid underground stem as compared with a bulb, which is a fleshy scaled leaf bud such as those produced by tulips and onions.

COROLLA. The inner series of floral leaves or petals which are the colored parts of most flowers.

CORYMB. A flat-topped or convex open flower cluster with stems arising at different levels. The marginal flowers open first.

CRYSTAL. A mineral in geometrical form with planes making up the outer surfaces. The internal arrangement of atoms determines the shape of the plane-sided structure.

CYME. A flat-topped flower cluster in which the central flowers bloom first.

DIASTROPHYSM. A term which encompasses all movements of the earth's crust, as in earthquakes, vulcanism, changes in position of rocks and also their deformation.

DIFFERENTIAL EROSION. Weathering of rocks at different rates, depending on the relative resistance or weakness of their substance to dessication by such factors as water, wind, or freezing and thawing.

DIKE. An intrusion of igneous rock into vertical cracks in bedrock. If the dike is of more resistant material than the surrounding rock, erosion may leave the dike exposed above the adjacent rocks. Dikes vary in thickness from inches to hundreds of feet and may be many miles in length.

DOME MOUNTAIN. A circular upfold of strata in which the rock layers dip downward in all directions from the center as in the Black Hills in South Dakota.

ECOSYSTEM. An ecological system or community in which producer, consumer, and decomposer organisms interact with each other and with inorganic and organic substances and the climate of the environment. The term includes environments such as a pond, a marsh with its surrounding feeding grounds, or a redwood forest.

EMERGENT PLANT. A plant which grows with some of its organs in water and others raised above the water.

ESCARPMENT. A long, linear rock exposure such as the surface of a cliff.

FALL LINE. The junction between the Atlantic Coastal Plain and the Piedmont of the older Appalachians. The name derives from the waterfalls in the rivers that flow from the harder rocks of the Appalachians onto the softer rocks of the Plain.

FAULT. A crack in the earth's crust along which rock movement has occurred.

FAULT-BLOCK MOUNTAIN (ALSO BLOCK MOUNTAIN). A mountain formed by uplift of a block of the earth's crust to form an escarpment on the faulted side and a long gentle slope on the side away from the fault. The Sierra Nevada and many of the mountains in the Basin and Range Province are block mountains.

FELDSPAR. A group of light colored aluminum silicate minerals found in igneous and metamorphic rocks. Feldspar contains differing amounts of sodium, potassium, and calcium. Some varieties occur as white or pink crystals.

FELLFIELD. An alpine community characterized by gravel and stones protruding through shallow soil where less than half the surface is covered with vegetation. Cushion plants are the predominant species on fellfields.

FOLD MOUNTAIN. A mountain formed by compression of the earth's crust into great folds such as occur in the Valley and Ridge Province in the Appalachian Highlands.

FOOD CHAIN. The transfer of energy from one organism to another through a series of steps as from plant, to rabbit, to coyote, to magpie or other scavenger. This is a predator chain. In a parasite chain a small organism, such as a bacterium or fungus, attacks a larger organism such as a deer or a tree. In a saprophytic chain microorganisms obtain food from nonliving organic matter. Interlocking food chains form a food web.

FORB. A term used to designate herbaceous plants, including useful plants as well as weeds, but not including grasses, sedges, shrubs, or trees.

FUMAROLE. A fissure in the rocks from which steam and other gases issue in volcanic areas such as Yellowstone and The Geysers, Sonoma County, California.

GALL. A swelling of plant tissue caused by insects which inject substances, either chemical or possibly bacteria, to produce a habitation and food for their larvae. The explanation of the abnormal plant growth is not clear. In some instances the larvae themselves may promote swelling of the plant tissues. Galls are common on roses, oaks, willows, and goldenrods.

GASTER. The enlarged part of the abdomen of an ant. The gaster contains the sex organs, various glands, and two "stomachs." The first stomach is actually a crop or elastic bag which carries food that can be regurgitated for larvae and other members of the colony. The second stomach is the ant's true stomach for digestive purposes of the insect itself.

GLACIER. A mass of ice formed from snow and frozen meltwater that moves downhill under the force of gravity. Continental glaciers covered almost a third of the earth's land surface during the Ice Age. Glaciers still occur in high elevation valleys in the northern mountains and especially in Alaska and Greenland.

GNEISS. A metamorphic rock with distinct light and dark layers or bands. Mica is one of the common constituents of gneiss (pronounced *nice*).

GRABEN. A valley caused by sinking of the earth's crust along fault lines.

HORNBLENDE. A mineral that occurs in long glossy black crystals and is found in many igneous and metamorphic rocks.

HORST. A massive block of the earth's crust lying between two faults and higher than the strata beyond the faults.

ICE AGE. The period of Pleistocene glaciation which began approximately two million years ago and ended 6,000 to 10,000 years ago. Ice sheets advanced and retreated four times during the Pleistocene. The periods between glacial advances are called interglacials. We are living now in a period which may be the fourth interglacial if the ice advances once more from the north.

IGNEOUS ROCK. Rock crystallized from a molten state. Molten magma erupted from a volcano and cooled is an example of igneous rock.

INVOLUCRE. A whorl of bracts or modified leaves surrounding a flower or flower cluster.

LAVA. The molten magma from a volcano. Also the crystallized rock when the magma cools.

LAVA BOMB. A lump of molten lava which cools in the air after being ejected from a volcano. These generally have a teardrop shape and may vary from an inch to several feet in length.

LOESS. Wind blown yellowish silt deposited to depths up to several hundred feet. Loess silts in the United States were blown out of Ice Age glacial deposits.

MAGMA. Molten silicate melt which cools to form igneous rock.

MESA. A hill with a flat, tablelike top. The top consists of harder material than the layers of softer rock below.

METAMORPHIC ROCK. Igneous or sedimentary rocks which have been changed by heat or pressure. The recrystallization produces crystals large enough to be seen, giving the rock a grainy appearance.

MICA. A mineral which occurs in thin, shiny, flat plates so soft they can be picked apart with a thumbnail. One form is known as isinglass. Mica is found in granite and other rocks.

MONADNOCK. Rock masses that stand alone on peneplains. They are remnants of rock layers that once extended to higher elevations.

MORAINE. The mixture of sand, silt, rocks and other materials transported by a glacier and deposited at its margin or at its lower end where the ice melts.

NICHE. The life style or behavior of an organism in a given habitat. The niche of an organism involves its activities in securing food, growing, reproducing, and contact and reaction to other organisms in its environment. Its habitat is the place where it lives.

OVIPOSIT. Egg laying by an insect by means of an ovipositor, or tubular structure, which may extend outside the abdomen and permit insertion of eggs into plants or the bodies of other organisms.

PAINT POT. Hot springs in which water bubbles or sputters through colored clays formed by weathering of volcanic rocks. Hot springs occur in Yellowstone and other volcanic regions.

PANICLE. An elongate branched inflorescence with the younger flowers at the top.

PEDICEL. The stalk of a single flower of an inflorescence.

PEDUNCLE. The stalk of a single flower or group of flowers of an inflorescence.

PENEPLAIN. An essentially level surface of a region formerly eroded to near sea level and subsequently elevated to a higher altitude.

PERIANTH. The calyx and corolla, especially when the two cannot be readily distinguished from one another.

PERIGYNIUM. The sac-like bract, sometimes inflated, that surrounds the pistil in the female flower of sedges.

PETIOLE. The stalk to a leaf blade.

PIKA. A small mammal of the alpine tundra, related to the rabbits and hares. It stores plants in "hay barns" in the rocks for its winter food. Also called cony.

PLANTIGRADE FEET. The feet of mammals, such as men, bears and shrews, which walk with the whole lower surface of the foot on the ground.

PLATEAU. A flat, elevated region, with essentially horizontal rock layers, often dissected by deep, steep-sided valleys.

PRIMARY CONSUMER. An organism which derives its food directly from plant material.

PUMICE. A light weight volcanic rock filled with air bubbles so as to resemble a sponge.

QUARTZ. A glassy mineral composed of silicon dioxide. Innumerable varieties exist: sand grains, agate, amethyst, and rock crystals.

RACEME. An elongated inflorescence with a single stem to which the flowers are attached by pedicels with the youngest flowers at the tip of the stem.

SANDSTONE. A sedimentary rock consisting of sand grains tightly cemented together.

SAPROPHYTE. A plant which derives its food substance from nonliving organic matter.

SCHIST. A metamorphic rock, fine-grained, with visible layers, and flaky when sufficient mica is present. Gneiss is similar but is coarser-grained.

SCREE. Loose rock debris that accumulates on talus slopes, generally in alpine areas. The rock particles commonly are gravel-sized. Talus usually consists of boulders and rocks down to fist-sized.

SEDIMENTARY ROCK. Rocks formed from material such as sand, mud, or gravel deposited by wind, water, or ice.

SEPALS. The outer whorl of the flower parts which are usually green as distinguished from the corolla which is colored.

SHALE. A finely layered sedimentary rock composed of clay minerals and usually compacted into sheets.

SOLIFLUCTION. A process whereby soil creeps downward on alpine slopes as a result of freezing and thawing above the bedrock or permafrost. Freezing and thawing creates solifluction terraces and frost hummocks.

SPERMATOPHORE. A mass of spermatozoa extruded by male salamanders and other primitive vertebrates which is inserted by the female into her cloaca to accomplish fertilization.

SPIKE. An elongated inflorescence with stalkless flowers. The younger flowers are at the tip.

STAMENS. The pollen-bearing organs of a flower. The stamen consists of a filament or stalk and an anther.

STIGMA. The part of the pistil of a flower that receives the pollen. The stigma usually is at the tip of the style or stalklike part of the pistil. The style connects the stigma with the ovary.

STRATA. Layers of sedimentary rock.

SUCCESSION. The sequence of biological communities which replace one another until a climax or steady state is reached.

TALUS SLOPE. An accumulation of rock fragments at the base of a cliff.

TARN. A small mountain lake.

THUNDER EGG. Local name for a geode which is a rounded concretionary rock, often hollow and lined with crystals or filled with minerals.

TUNDRA. The arctic grassland consisting of grasses, sedges, and dwarf shrubs which grow on a spongy mat of living and undecayed vegetation. The climate is characterized by low temperature and a short growing season.

WATER GAP. A narrow valley cut through a mountain ridge by a river which flows from one side of the mountains to the other.

Water gaps permit passage of railroads and automobile high-ways through instead of over the mountains. The Delaware Water Gap, cut by the Delaware River through the Kittatinny Mountains in Pennsylvania is a famous one.

WIND GAP. A former water gap where stream piracy has diverted the original river from crossing the mountains. These are common in the southern Appalachians.

Bibliography

ADAMS, KRAMER. A. *Logging Railroads of the West*. New York: Bonanza Books, 1961.

ALEXANDER, G. "The occurrence of Orthoptera at high altitudes with special reference to Colorado Acrididae." *Ecology*, 32: 104-112, 1951.

ANDERSON, ALAN H., JR. *The Drifting Continents*. New York: G. P. Putnam's Sons, 1971.

APPALACHIAN MOUNTAIN CLUB. *Mountain Flowers of New England*. Boston, 1964.

BAARS, DONALD L. *Red Rock Country. The Geological History of the Colorado Plateau*. Garden City, New York: Natural History Press (Doubleday), 1972.

BARKER, W. "Haymaker of the high spots." *Natural History*, 65: 46-47, 1956.

BAXTER, PERCIVAL PROCTOR, CONSTANCE BAXTER, JUDITH A. AND JOHN W. HAKOLA. *Greatest Mountain: Katahdin's Wilderness*. Centerville, Maine: Scrimshaw Press, 1972.

BEEBE, LUCIUS, AND CHARLES CLEGG. *Narrow Gauge in the Rockies*. Berkeley: Howell-North Books, 1958.

BERGLUND, BERNDT. *Wilderness Survival. A Complete Handbook and Guide for Survival in the North American Wilds*. New York: Charles Scribner's Sons, 1972.

BILLINGS, W. D. "Arctic and alpine vegetations: similarities, differences, and susceptibility to disturbance." *Bio Science*, 23:697-704, 1973.

———— AND L. C. BLISS. "An alpine snowbank environment and its effects on vegetation, plant development, and productivity." *Ecology*, 40:388-397, 1959.

BLISS, L. C. "Alpine plant communities of the Presidential Range, New Hampshire." *Ecology*, 44:678-697, 1963.

BOLLINGER, EDWARD T., AND FREDERICK BAUER. *The Moffat Road.* Denver: Sage Books, 1962.

BROOKS, MAURICE. *The Appalachians.* Boston: Houghton Mifflin Company, 1965.

————. *The Life of the Mountains.* New York: McGraw-Hill Book Company, 1967.

BROWN, F. M. "Colorado butterflies." *Proceedings of the Denver Museum of Natural History*, Nos. 3-7, 1957.

BROWN, ROBERT L. *Colorado Ghost Towns—Past and Present.* Caldwell, Idaho: The Caxton Printers, Ltd., 1972.

————. *An Empire of Silver. A History of the San Juan Silver Rush.* Caldwell, Idaho: The Caxton Printers, Ltd., 1965.

————. *Jeep Trails to Colorado Ghost Towns.* Caldwell, Idaho: The Caxton Printers, Ltd., 1964.

CHAPMAN, J. A. "Studies on summit-frequenting insects in western Montana." *Ecology*, 35:41-49, 1954.

CRANDELL, DWIGHT R. "The geologic story of Mount Rainier." U.S. Department of the Interior, Geological Survey Bulletin 1292, 1969.

DAVIES, J. L. *Landforms of Cold Climates.* Cambridge, Mass.: M.I.T. Press, 1972.

EBERHART, PERRY. *Guide to the Colorado Ghost Towns and Mining Camps.* Denver: Sage Books, 1959.

———— AND PHILIP SCHMUCK. *The Fourteeners—Colorado's Great Mountains.* Chicago: Sage Books, The Swallow Press, Inc., 1970.

EKMAN, LEONARD C. *Scenic Geology of the Pacific Northwest.* Portland, Oregon: Binfords & Mort, Publisher, 1970.

EMBLETON, CLIFFORD (ED.). *Glaciers and Glacial Erosion.* New York: Macmillan Publishing Co., 1973.

FIELDER, MILDRED. *Railroads of the Black Hills.* New York: Bonanza Books, 1964.

FISHER, RONALD M. *The Appalachian Trail.* Washington: National Geographic Society, 1972.

FLEMING, HOWARD. *Narrow Gauge Railways in America.* Oakland: Grahame Hardy, 1949.

GOLDEN, FREDERIC. *The Moving Continents.* New York: Charles Scribner's Sons, 1972.

GREEN MOUNTAIN CLUB. *Guide Book of the Long Trail.* Rutland, Vermont: Green Mountain Club, 1970.

GRIGGS, R. F. "The timberlines of northern America and their interpretation." *Ecology*, 27:275-289, 1946.

HALLAM, A. *A Revolution in the Earth Sciences. From Continental Drift to Plate Tectonics.* New York: Oxford University Press, 1973.

HARWOOD, MICHAEL. *The View from Hawk Mountain.* New York: Charles Scribner's Sons, 1973.

HAYWARD, C. L. "Alpine biotic communities of the Uinta Mountains, Utah." *Ecological Monographs*, 22:93-120, 1952.

HEINOLD, GEORGE. "It's no disgrace to be a rat." *The Saturday Evening Post*, Nov. 8, 1947, pp. 30, 157-158.

HUBLEY, R. C. "Glaciers of the Washington Cascade and Olympic Mountains; their present activity and its relation to local climatic trends." *Journal of Glaciology*, 2:669-683, 1956.

HUDOWALSKI, GRACE L. (ED.). *The Adirondack High Peaks and the Forty-sixers.* Albany, New York: The Peters Print, 1971.

HUHEEY, JAMES E., AND ARTHUR STUPKA. *Amphibians and Reptiles of Great Smoky Mountains National Park.* Knoxville: University of Tennessee Press, 1967.

IVES, R. L. "The beaver-meadow complex." *Journal of Geomorphology*, 5:191-203, 1942.

JOHNSTON, VERNA R. *Sierra Nevada.* Boston: Houghton Mifflin Company, 1973.

LECHLEITNER, R. R. *Wild Mammals of Colorado.* Boulder: Pruett Publishing Company, 1969.

LITTLE, E. L. "Alpine flora of San Francisco Mountain, Arizona." *Madroño*, 6:65-81, 1941.

LLOYD, ROBERT M., AND RICHARD S. MITCHELL. *A Flora of the White Mountains, California and Nevada.* Berkeley: University of California Press, 1973.

McINTYRE, MICHAEL P. *Physical Geography.* New York: Ronald Press, 1973.

MILNE, LORUS J., MARGERY MILNE, AND THE EDITORS OF *Life. The Mountains.* New York: Life Nature Library, Time Incorporated, 1962.

MOORE, DWIGHT M. *Trees of Arkansas* (3rd revised edition). Little Rock: Arkansas Forestry Commission, 1972.

MUNZ, PHILIP A. *California Mountain Wildflowers.* Berkeley: University of California Press, 1963.

OSBURN, W. S. "Characteristics of the *Kobresia bellardi* meadow stand ecosystem in the Front Range, Colorado." *Journal Colorado-Wyoming Academy of Science,* 4:38-39, 1958.

SAUNDERS, ARETAS A. "The summer birds of the northern Adirondack mountains." *Roosevelt Wild Life Bulletin,* 5:327-504, 1929.

SCHLEGEL, DOROTHY M. "Gem stones of the United States." United States Department of the Interior. *Geological Survey Bulletin* 1042-G.

SCHWAN, H. E., AND D. F. COSTELLO. *The Rocky Mountain Alpine Type—Range Conditions, Trends and Land Use.* Denver: U.S. Forest Service, 1951.

SHARPE, GRANT W. *101 Wildflowers of Glacier National Park.* Glacier Natural History Association in Cooperation with the National Park Service, U.S. Department of the Interior, Special Bulletin No. 5, 1967.

SPRAGUE, MARSHALL. *Great Gates: The Story of the Rocky Mountain Passes.* Boston: Little, Brown & Company, 1964.

STEBBINS, ROBERT C. *Amphibians and Reptiles of California.* Berkeley: University of California Press, 1972.

STORER, TRACY I., AND USINGER, ROBERT L. *Sierra Nevada Natural History: An Illustrated Handbook.* Berkeley: University of California Press, 1963.

STUPKA, ARTHUR. *Wildflowers in Color* (Eastern Edition). New York: Harper & Row, Publishers, 1965.

U.S. DEPARTMENT OF THE INTERIOR, GEOLOGICAL SURVEY. "Elevations and Distances in the United States." Washington, D.C.: Government Printing Office. USGS, INF-73-7.

WEIS, NORMAN D. *Ghost Towns of the Northwest.* Caldwell, Idaho: The Caxton Printers, Ltd., 1971.

ZWINGER, ANN H., AND BEATRICE E. WILLARD. *Land above the Trees. A Guide to American Alpine Tundra.* New York: Harper and Row, Publishers, 1972.

Index